BLUE JEANS TO OLIVE GREENS

Meet Jay, an army brat, whose sole aim, to have fun, is in direct conflict with his father's aim for Jay, to get into the Army. Coming from a family with three generations of proud soldiers can be such a drag! The only way to get away and get girls is to join the family business...or so he thinks.

Follow the young hero as he makes his way from the comfort of his home to the harsh reality of a trainee officer.

Written from a personal viewpoint, by a fourth generation Army officer for whom the army not a job but a way of life, and not seriousness either.

Blue Jeans to Olive Greens is a fun, humorous and enlightening read that takes you into the hallowed portals of one of the most distinguished training institutions in the country, The IMA. Be astonished, amused and inspired by the hard road that Jay needs to walk on to become the best kind of hero, a soldier.

BLUE JEANS TO OLIVE GREENS

VINOD BHASKER

Foreword by

Lt Gen Raj Kadyan (Retd)

PENTAGON PRESS

Blue Jeans to Olive Greens
Lt Col Vinod Bhasker

First Published in 2014

ISBN 978-81-8274-802-6

Published by
PENTAGON PRESS
206, Peacock Lane, Shahpur Jat,
New Delhi-110049
Phones: 011-64706243, 26491568
Telefax: 011-26490600
email: rajan@pentagonpress.in
website: www.pentagonpress.in

Branch:
Prime Arcade
Office #11
1154 Saifee Street
Opp. M.G.Road, Camp
Pune-411001
Email: pentagonpresspune@gmail.com

Printed at Aegean Offset Printers, Greater Noida, U.P.

CONTENTS

BOOK III
MAN TO SOLDIER - II TERM

FOREWORD

It gives me great pleasure to write a foreword for the book written by Vinod Bhasker. More so since Vinod's father Brigadier Henry Bhasker and I carry the same bottling label of Army life. Besides being batch-mates, we also commanded neigbouring posts in 1966 along the very active Cease Fire Line in the Rajouri Sector. It thus gives me some kind of avuncular pride in penning these lines.

The book describes the author's early Army experience including the first grounding in the Indian Military Academy. The IMA is a foundry that moulds a carefree college student into a disciplined individuals in a short span of eighteen months. It tames the 'tigers' into disciplined men that would lead their troops in trying and dangerous situations. The process of change is harsh but has its lighter sides. The story of the British drill Sergeant Major, who trained the first few post-Independence batches typifies the rustic humour of the uniformed men. On a Monday morning parade when he noticed that one of the gentlemen cadets had not taken the regulation weekly haircut on the preceding day, he stood behind the offender and asked, "Am I hurting you sir?" "No sir" the cadet shouted his reply. "I thought I may be standing on your hair sir" was the Sergeant Major's wry response. As time passes one begins to enjoy the 'suffering'. Peer strength is one of the most effective ameliorative factors. You become a part of the family and share your thoughts and burdens freely. These get laughed off.

Average human life is full of stress. Work pressure, criticism, fatigue, boredom, rejection by a loved one, all contribute to the stress. Add to these the tough training and gruelling schedule necessary before one can qualify for being commissioned as an officer. The best way to bear the stress is to start enjoying it. And that is where humour comes in. Humour gives harmless pleasure and is a great buffer against stress. One does not have to be a specialist to find humour. It is available in everyday life and commonplace situations. One just has to have an observant eye to discover it. Military life with such a variant landscape of situations is a fertile source of humour. The hard life has an hidden underbelly that can tickle the toughest bone. Military environment is peculiarly suited to humour. In fact 'Humour in uniform' is a well-known subject by itself. Humour is not so much about creating loud guffaws of laughter or any intellectual exercise. Humour is more about spreading smiles by seeing through and describing the lighter side of life. Humour deals with people and their oddities; it is observational. And that is what the author has attempted to highlight in his aptly titled book 'From Blue Jeans to Olive Greens'. The book is based on the author's early days in uniform including life at the IMA. The experience varies as one acquires a stronger tan of the olive green. First termers in the IMA are the green horns, bossed and bullied, even berated. All 'yes sir' mostly to those that have just graduated to the second term and feel senior for the first time. The military reader will not fail to recall his own days when as freshers we ran in our newly issued nailed boots with blistered feet and hurting toe nails, but still laughing at our plight.

Spirit of humour seems to alight on the author quite unbidden. He has a natural flair for seeing the lighter side of life and finding appropriate words to describe it. He has the ability to pick an odd, embarrassing or funny incident experienced or observed and describe it in a plain manner. His style is a mix of fun and a sense of proportion. He has cobbled together the detached fragments of his early experience in the olive greens in an enjoyable prose. Military readers no doubt would be able to relate to the situations more easily, but others will like it equally.

'Blue jeans to olive greens' is an enjoyable read.

Gurgaon (Haryana) **Lt Gen Raj Kadyan (Retd)**

PREFACE

'A preface generally covers the story of how the book came into being, or how the idea for the book was developed; this is often followed by thanks and acknowledgments to people who were helpful to the author during the time of writing.' At least that's what Wikipedia says and I intend to abide by it.

The idea for this book came about many years ago, when I read a British officer's light-hearted account of his life in the Army. The experiences he described occurred nearly a hundred years ago but were still quite relevant and interesting. He went on to become a successful, and more importantly, a rich novelist and I, a poor army officer. My journey on the path of fiction actually started out with my telling funny and wildly imaginative stories to my parents. It was usually to escape punishment for my many transgressions.

After joining the Army, I wrote articles in a humorous vein and was given to believe that that was my forte. I found it easier to make people laugh, though officers in any hierarchical organization usually find it easier to make people cry. I racked my brains for a non-controversial topic and decided to write about my cadet days. Most creatures are nostalgic about their younger days. I also did not want to upset anyone and lose my pension, the latter being more important.

The Indian Military Academy or IMA is a wonderful establishment. It takes in men; actually some of them still boys, from extremely diverse backgrounds and races, and turns them into leaders with almost identical mindsets and values. Not an easy task to do in just 18 months; in effect undoing something which had probably taken about 18 years of parental dedication and societal pressure.

Life in the Academy is tough and at times brutal. But my experience in the Army, in fact in life, is that light moments can be found no matter how grave the situation. The events described are based mostly on my own or other's experiences and stories or legends that I have heard. The idea was never to denigrate any class or community. It would be pertinent to mention here that in spite of the politically incorrect language that is commonly used in the Army it is almost never taken amiss. I have rarely heard any references to caste or religion. I myself have been called '*kalu*' or 'blackie' as my father before me, while in the regiment. We never minded...our wives might have, though.

Talking about language, men especially in the Army, when not in the presence of ladies, tend to swear a lot. I have attempted to recreate the general atmosphere without being too graphic as some of the actual dialogue was, at times. If anyone says otherwise I would like to definitely meet that person and or group, as I have never had the privilege of meeting aliens before.

On that note I must tell you, my father read a chapter and said tersely, "Where the f@# did you learn to use such bloody language!" My mother is still praying for my soul after reading just a few pages. My gentle wife read the synopsis and got a headache. But it was thanks to the encouragement of my friends, Srividya who had the onerous task of editing the book and Shantini, the only person who has consistently found the book funny, that it has seen the light of the day or life outside my laptop. Of course, the support of my family, especially my wife and kids who keep asking when the money is going to start rolling in and who I like to think believed in me, made me start and keep on writing. Finally, one cannot miss mentioning my regiment and *paltan*s, 12th and 18th battalions of The Mahar Regiment with whom I have spent the best and most inspiring years of my life, and made me a decent human being...I think. They are still my home, and they will always have my back. I would also like to thank my senior subaltern Col RS Raman for the cartoons and his sense of humor that has figured in this book of fiction and rescued me from boredom in real life.

Lastly, no disrespect is meant to civilians and none were seriously injured or insulted in the making of this book. My wife and mother are civilians and proud of it, and this book was edited by a civilian. The term 'civilian' has been used historically and currently by armies across the globe to denote someone without training or discipline...among other things. It is not my invention, it's just tradition and I hope lots of civilians read my books. Any other hurt feelings are just collateral damage and regretted. Anyone feeling very strongly may join me for a drink for the traditional therapy for hurt feelings. This book is based on my memories which are somewhat frozen in the 90's. I'm sure things have changed and improved as they should, since. The sequences of events have been changed and I have given myself fictional license, mostly so the characters and stories are not easily recognizable. I'm sure my dear coursemates and friends will forgive me. I may not have been able to add every facet of academy life but will try and do so, as far as possible, in the sequel to this book. Some may feel I have added unnecessary explanatory notes. But that is only on the request of some of my civilian and foreign friends who found some terms and aspects quite alien.

VINOD BHASKER

BOOK I

THE BOY

Chapter 1

LOVE OVER DUTY

The shopping centre next to the park was calm the crowd disciplined. Like in any other military station, the evening was peaceful as people wound down after a hectic day. But calm was not the state that could describe Jay Shankar's mind or orderly his state. All he could think about was the upcoming party and the girls who would be attending it.

The 'party' was also enthusiastically referred to as a 'jam session'. It was cantonment slang for a dance party. It was actually more of a teen mating ritual without the obvious conclusion. It was hopes of a conclusion of some sort that had Jay in a tizzy.

The 'jam session' had Western intentions but mostly disappointing Indian conclusions.

Jay and his gang had been looking forward to it for some time now. The term 'gang' was misleading; it referred to a group of scruffy, horny yet well-mannered teenagers. Rat pack would have been more apt.

The master and mistress of the house where the party was being held were out of station. This heightened Jay's anticipation and hormone levels. He was eagerly looking forward to a heady combination of wine, women and song.

These were things that mature men did not pass up easily and our hero was but a sixteen year old. He, sadly and not unsurprisingly, did not have a girlfriend. This was partially because of his lack of confidence and no lack of pimples. These slurs Jay and his loving mother would have hotly contested, and his hostile younger sister eagerly confirmed.

Jay's situation was not unique. Such was the social norm in any Indian military colony or as it was popularly called cantonment or cantt. for short, which was also the official abbreviation. But there was another somewhat official reason for the pressure.

You see, Jay had the NDA (National Defence Academy) exams coming up and he was supposed to be studying, studying very hard indeed. This put a lot of pressure on young Jay. He was torn between the prospect of love and the exhortations of duty. Those 'calls to duty' his father regularly threw his way, at least more regularly than pocket money.

The singular reality that had penetrated his psyche, through daily doses of parental sermons, was simple. He would not be able to attract or buy love till he did his duty. That he would do by getting a job, respect and money, strictly in that order. The only one that could guarantee that was the Army or so he was informed, at least twice a day... every day.

"It's all about respect, not bloody money," thundered Jay's father, very often. "Unfortunately not much respect or enough money," Jay's mother usually added in an undertone.

The duty factor was a result of his father's solemn directions to join the 'family business' by getting through the NDA entrance exams. After all his was to be the fourth generation, hopefully, to join the Army-beginning with his great grandfather.

Great *dadaji* had been a *havildar* in the Great War and his son a *jemadar* or VCO, a viceroy commissioned officer in World War II. Dad had become a colonel with vague career prospects and felt strongly that it was Master Jay Shankar's responsibility to produce at least one general in the family.

To Jay's hormone driven mind, being practical or having practical knowledge was not a priority. He would think about the blessed NDA entrance exams a little later. Right now it was Jaswinder, that sexy young newcomer to the cantt. who was on his mind. Rumour had it that her father had spent a few years in Italy as a defence attaché and she was not *really* a virgin. That Italian connection, though irrelevant, gave Jay the shivers of anticipation.

Though if questioned he would not have been able to define what he was anticipating. But it was enough to ensure that thoughts of the impending exams receded.

If his father but knew, that it was not the NDA that was uppermost on his agenda, life under a '*kadak fauji baap*' could get real bad. But Jay didn't really care. He was being controlled not by rationale and his brain, but by impure thoughts and other organs.

Jay then turned his thoughts to that ass, Sharma, just back from his first term in the NDA, behaving like he was some bloody war veteran. The height of pretentiousness was Cadet Nitin Sharma of Lima Squadron, going with his adoring mom for walks, dressed in his NDA muftis aka tie, blazer; generally exuding the aura of a soldier, ready to be called up for duty any time.

The ass had been one of the first from the cantt. to join the NDA, and now everyone was treating him like he was some damn war hero. Aunties everywhere would gush and tell Mrs Sharma, "How proud you must be to have a son who was 'settled' so early!" Mummy Sharma lapped it all up with a superior smile. The inference being her womb had produced junior Sharma, and thus been welcomed to the Army. Her smugness also indicated that in due course her progeny would become a general or some other worthy of national importance.

Not to be outdone by his wife, Senior (Col) Sharma would make it a point to hang out with his son at key points during club events. Especially points like the bar or bathroom, where the public and especially some general would be sure to pass through or by. He would then lap up the VIPs' inevitable felicitations. There had been this embarrassing moment when Jay's father had bumped into Sharma*ji* at the same ambush point. Having no other choice, Shankar senior had congratulated Sharma senior. The greetings were cordial though both seniors despised each other.

Senior Sharma had actually sneered and asked Jay's father if Jay was appearing for the NDA exam or.... The 'or' inference would get under the skin of any self-respecting parent. Colonel Vinay Shankar, Jay's father smiled, gritted his teeth and replied that Jay would probably be trying for IMA after graduation. The reason given, though in a thoroughly unconvincing tone, "I want my son to see some life before getting into the Army".

But Jay was not fooled by this catty exchange; he knew what awaited him once back home. The whole drama was quite predictable every time those silly NDA buggers came home on leave.

The moment they got home his father would open the floor with, "What the hell are you doing just lying about the house like a useless rag?!"

Jay's reply was equally predictable, "What would you have me do? I've not yet finished high school."

"Study... do something, prepare for NDA at least...see those Sharma and Mehta boys—they looks so smart with their haircuts and scarves". This dialogue included a larger and captive audience, comprising Jay's sister and mother.

They were not terribly appreciative of either the senior's jibes or junior's cheek. But being women and more pragmatic they knew better than to openly take sides. They knew it was no contest between a penniless and powerless loser versus the breadwinner.

Jay found this lack of support very hurtful and saw it as a total betrayal. He had already decided what gifts he would NOT be getting them when he started earning, someday.

Jay's rebuttal for the defence was equally predictable: "Mehta looks like a moron with that stupid zero cut...looks like someone put a bowl on his head in an assembly line and used a machine below the bowl line!" he took a deep breath and delivered the final grenade in his arsenal. "That idiot doesn't even know what Pink Floyd means...he thinks it's some cocktail".

This was pretty close to the truth and his dad would mutter something like, "Nonsense he looks fit and how can you call him a moron... he got through the NDA exam, didn't he?!" The 'unlike you' had originally formed part of the sentence but was now left unsaid for socio-political reasons. Jay's mother might not have liked that slur on her genes or progeny.

Jay was just an average student and reasonably good at games. These were mostly 'officer like' games, namely tennis, squash and a bit of golf. He was like any other average Army officer's kid. His mother, like all mothers, civilian or Army, believed he was God's gift to Earth. The odious comparisons, especially to dopes like Cadet Sharma and Cadet Corporal Harish Mehta were downright insulting. They were also average but had now been elevated to 'outstanding' grading after joining NDA.

Cadet Harish Mehta had been a bore before, and he was a bigger one now... always yapping about his experiences in the academy. The cross-country run tale was described in tones normally reserved for recalling desperate and bloody battles. But he was on the approved list of boys Jay could hang out with so he had no choice but to listen to the monologues.

He recalled their last encounter at his friend Raj's place. Rajeshwar Ahlawat was also desperate about joining the Army. His father too was 'after his blood'. The fact that Raj's elder brother was about to pass out from the IMA did not help the situation. Raj's parents had decided that Mehta was the ideal role model. Raj had been instructed to meet Mehta and get tips on how to ace the NDA exam. Raj had called Jay for support through the ordeal.

Jay knew Mehta well and was not very hopeful of getting much by way of

NDA exam tips, but it was his 'duty' to help his friend. Wasn't that what soldiers did? There was the 'duty' word again, a result of growing up in an environment where everyone talked about duty or going on duty.

Mehta never stopped talking about all the punishment parades and cross-country practice...yes he had actually run seven whole kilometres. Mehta 'sir' was trying to motivate people to join the NDA by relating the horrors. He wasn't doing a great job.

Raj and Jay had been suitably awed and worried by the seven-kilometre thing. After all they had never walked that much before, leave alone run. The eager beavers then asked the question that had been nagging them all along, "*Yaar, humne suna hai* there is a lot of *homopanthi* in your squadron."

Mehta's reply shook them to the core, "*Nahin yaar*! It happens sometimes only...in the heat of the moment!!" Jay and Raj looked at each other with slack jaws. They could not believe what they had just heard. Their eyes met and they shared the same unspoken fear of being violated in some 'heated moment'. There seemed to be too many of those sorts of moments in military circles, where tempers frequently ran high.

The more they listened to Mehta, the more they had a sneaking suspicion that he was exaggerating. It was humanly impossible to pack in so much adventure and life experience into just eighteen odd months. Mehta the Rambo had just graduated into his third term. But even after three hours, Mehta had given them precious little by way of concrete tips on how to get into the NDA.

The bottom line was work hard, study everything, leave nothing to chance, start preparing at least one year in advance and lastly, have extra coaching during the holidays. This same homily was usually written on the NDA guidebook introduction or preface, and was not the outcome of any significant sagacity. Besides, they heard different versions and words to that effect from their parents nearly every day...twice a day on an average.

It occurred to the two army wannabes that this advice required a very high degree of sincerity and dedication. Something they were not sure they had, but it explained why Mehta was almost brain dead.

They reached home still debating the merits of joining the army. Raj's mother opened the door and had the last word as she shooed Jay away, "*Beta* don't distract him, *use padhne de* NDA *ke liye.*" In other words, "Don't come here again you pest, you are the prime hurdle in our boy's future plans".

Jay gulped in embarrassment and left. He read the writing on the wall as the door slammed in his face, again. He sincerely wished Raj's parents luck with their efforts in getting their moron son and his closest friend recruited into NDA.

Jay knew there was not much of a point in calling Raj up right then and checking if he would be at the jam session. The more he thought about it, the stickier his attendance at the jam session looked. The NDA exam was but three days away and everyone else would be cramming seriously.

It felt kind of embarrassing to be at the party while other kids were cramming seriously. That sort of thing would invariably reach his father's ears. Jay's dutiful dad did not take such 'desertions from duty' very kindly.

He wondered if a cigarette would help him think clearly, but discarded that idea. The last time he had tried smoking his mother had smelt the smoke and slapped him silly. Strangely his father had not said much, probably because he had quit smoking only recently.

The next evening Jay told his father that he was going to Raj's place to study. He ignored his sister's smirk and was about to escape when he heard his sister say, "Don't you think he's a bit too well dressed for just a visit to Raj?"

Jay's heart skipped a beat. The repercussions, should his perfidy be discovered, would be swift and harsh. Jay almost abandoned his plans, but took heart when he heard his mother murmur "Must be to impress Raj's sister..."

Jay removed himself from the scene rapidly before any further suspicions were aroused. He went to Raj's house, sneaked around to the back and knocked on his window. Raj opened up, and asked eagerly, "Jam session?!" The answer was an emphatic, "Hell yeah!"

Having ascertained Raj's intentions, Jay went back and knocked on the front door and informed Raj's mother that they were going to Rajesh's place for some group study. The lady beamed at this proof of noble and high thinking and gave her blessings. Five minutes later, the two were on their way to the jam session.

Rajesh was another NDA hopeful; at least his parents were hopeful. His peers had no such high hopes for Rajesh's ability to become anything more than a peon. He had a good physique and very little intelligence. He had survived thus far through luck and the good offices of a very beautiful sister.

Rajesh was not the most reliable alibi since he could be trusted to screw up the simplest instructions. Jay decided to call and tell him anyway since they were all out of options. Rajesh would not be coming for the jam session, it was highly unlikely that the dragon he called 'mom' would let him. Jay hated having to speak to her but the call had to be made if their alibi had to be maintained. The conversation was as expected, civil and unhappy for both parties.

"Good evening aunty. May I speak to Rajesh?"

"*Beta,* you know he is studying for the NDA exam!"

"Uhh... aunty actually I wanted to clarify some doubts..."

"Ok, but make it short he has to study, so please don't disturb him time and again!"

She also added 'useless child' in an undertone which Jay was sure to hear. After an interminable wait, Rajesh came on the line and Jay told him to tell anyone who asked, especially Jay's parents, that they were in an intense group study session at his place. He quickly hung up before Rajesh asked any more questions and confused himself and others. The boy was talented that way.

The music at the jam session was good and so was the crowd. One of the guys had sneaked in some booze from home and it acted as a 'booster charge'.[1] It would have been more fun if there had been more girls without brothers, particularly younger brothers. Those insects understood their importance but overestimated their cuteness. These younger brother types had to be borne and tolerated if one wished to retain some vestigial interest in the eyes of those sexy sisters. The guys even had a name for these creatures, 'BOSS', short for 'brother of sexy sister'.

The cadets, Mehta and Sharma, were there wearing smart shirts and pants unlike the jeans and sweatshirts that the rest of the 'civilians' sported. The cadets loved to refer to the rest of humanity as civilians. Not wearing ties and blazers was as close as they could come to letting their hair down. They looked down with disdain at the undisciplined and degenerate youth around them. They sneered at the crowd and generally tried to stay aloof as the other kids drank, danced and made eyes at the opposite sex.

After a while, the cadets condescended to really let their (crew cut) hair down. This the dynamic duo did by commencing enthusiastic and erotic stag dancing, till they were sweating copiously. The crowd was impressed by their stamina and more than a little touched by the unnatural dance moves of the alpha and bravo military males (they had already established senior junior positions).

Raj and Jay concluded that cross-country may help you dance all night with a lascivious vigour. It seemed kind of sad that the NDA guys seemed to be happier dancing with each other while the girls admired their marriage potential. The fact that there were two dedicated teenaged romantics available and waiting in the wings did not lessen the injustice of it all. They remembered Mehta's lame rebuttal of the rumours of homosexual activity. They looked at the girls and at each other and shook their heads at the injustice of *karma*.

Raj and Jay finally managed to get up the courage and ask for a dance before it got too late. Their courage was desperate, strained by time, feminine reticence and paternal terror. Anytime now parents would start wondering if their kids were really out studying. Jay stalked Jaswinder till she was alone and popped the question. She said 'yes' and they danced. Jay may as well have been dancing with her chest since he couldn't take his eyes off it. Luckily she did not notice in the dim lighting. She was quite dim herself.

Jay danced like it was his last time. There were many things that could happen, namely, Rajesh's mother calling up, Rajesh coming to their home to meet them or their sisters sneaking the news of a jam session. Then it would probably be the last time he danced before he got a job or got kicked out of the house, whichever came first.

These were some of the 'most likely' and 'likely to happen any time' situations. Jay and Raj could only hope that they made it home before these things happened. But right now the sight of their young partners and their jiggling extremities had them mesmerised. It was with great regret that they broke the trance and took their leave of the party. They had to leave early in the morning for Chandigarh, the nearest exam centre.

As they made their exit they were met by Rajesh. "Hey guys, there you are. I went looking for you chaps and your sister told us you might be here..." said Rajesh cheerily to Raj. Raj rushed off into the dark with Jay behind him.

The next day there were about twenty of them in the back of the uncomfortable Army truck which smelt of stale milk and vegetables. Jay had a hangover, thanks to that silly Raj who had smuggled rum from his father's stock in a shampoo bottle. Jay could bet the idiot had not rinsed the bottle properly. And now his gastric system was adding to the atmosphere, though thankfully muffled by the chorus of the truck's diesel engine.

Jay tried to ease his discomfort by thinking about the new girl he had danced with and the way her boobs had jiggled. He felt someone jiggling next to him—it was a big fat, teenaged, pimply *sardar* or 'surdie'. He was the nerdy, studious type unlike most surds. He was still big enough to be menacing anyway, like most of his clan. Jay knew the type and tried to ignore his blubber jiggling his elbow.

The nerdy surdie boy kept yakking about how he had prepared for the test and how he was sure to get through, and dad would get him a new Yamaha...blah, blah, blah. He looked for a sign from God to relieve him of his torment especially the one his neighbour was inflicting on him.

Suddenly surdie switched tracks, "How was the party last night?"

Jay was a bit taken aback, "How the hell did you get to know?"

"My sister told me...she was there," said surdie with a wide anticipatory grin, "So how was it?"

Jay realised that if the sisters knew, his parents would too. But now it was time to get his own back at the surdie... by making him envious of what he had missed at the jam session.

He said, "It was fun *yaar* and there was this babe who was really hot. I had fun dancing with her. She was also game. When we get back I'm going to..." and Jay launched into a vividly graphic description that was literally too good to be true morally and physically. But it soon had surdie's glasses all fogged up.

The description was based on a page from a *Playboy* magazine given by a young artillery captain. The bribe was given in the mistaken hopes of getting an introduction to Jay's sister. It was a case of mistaken identity; the arty guy had actually been after someone else's sister. But the brief con had netted enough porn and that now provided Jay an opportunity to relate passages verbatim. His fantastic narrative had his new found friend enthralled and drooling.

After a while surdie could bear it no longer and asked, "What was the girl's name *yaar*?"

Jay said casually, "Can't really remember, but she has really big tits and lives at 17 Cariappa Marg..."

The surdie looked at him intently, then got up and changed his seat while the *danda*man[2] kept screaming at the surdie to sit down. Surdie kept giving Jay venomous looks from his corner. Jay asked Raj, "What the hell is the matter with that fat surd?"

Raj laughed, "You sorry bum, that was his sister you were talking about..."

Jay sighed heavily and said, "Oh shit!" The expletive covered a range of emotions none of which was shame. 'Guess jiggling flesh runs in the family' thought Jay and tried to cheer himself up with that thought.

He knew he had a very short reprieve. Fat nerdy surdie would surely find him and beat the shit out of him after the CDS (Combined Defence Services) exam. What was worse was that he was sure fatso would get into the Army before him and would be waiting at the Academy...

NOTES

1. Booster charge—used in mortar bombs to increase range, commonly used as slang for increased performance or kick.
2. Dandaman - The person depicted to sit in the back of a military truck and bang the tail board with a 'danda' on stick wherever a stop on emergency occurred.

Chapter 2

THE END OF FREE FOOD

Jay sighed as he looked at the bundles of Competition Success Reviews and Competition Master Magazines. They were one year's worth of back issues with tons of old question papers and other bits of information thrown in as well. Some of the info was useful, others supposedly inspirational, the balance depressing. There were interviews and stories about toppers in various entrance exams for hallowed institutions like the IIT (Indian Institute of Technology), Indian Civil Services. These were the holy grail of all job seekers, and even medical and engineering institute aspirants. The military was a consolation prize.

He wasn't very impressed with the pudgy nerds who appeared as success stories. More often than not the pictures showed kids with signs of 'puja' with vermillion or other tic-tac-toe type marks on the victim's forehead...Probably to propitiate the gods for continued success and bigger bribes.

Jay's parents were very impressed too. They hoped to see Jay's picture there someday. Filial loyalty is quite blind and a great believer in miracles. Jay's experience with the NDA exams had not dampened their ardour to see him in the family business. They were sure he would get through to IMA or at the very least, OTA[1] . He was in the final year of college and his father's ultimatums had taken on a tone of finality.

But Jay had not seen any interviews or pictures of CDS (Combined Defence Services exam) toppers in any magazines. Maybe they weren't considered worthy role models...a fact he had tried to point out to the head of the family only to be brusquely shut up. Maybe the old man had a point; unlike other exams, the written exam results of the CDS alone were not what decided the merit for the NDA or IMA.

There were the multiple choice papers followed by the SSB or the Services Selection Board, a four-day long selection procedure and only if you got through that, would you go for the stringent medical evaluation. The marks one got in the SSB would be combined with the written test marks to make the final merit, with the SSB marks usually making the difference between IMA and OTA. It was the other way around with the civil services where the nerds ruled.

All this was very daunting as were the stacks of material full of useless facts he had to memorize and of course, mathematics, his Waterloo, which he had to master. The facts he had to memorise were basically frustrating trivia like where was the maximum copper found in India. He was trying to become an officer for God's sake, not a prospector.

Col Shankar swore he had all these facts at his fingertips thanks to his habit of reading newspapers daily. The only thing Jay read the papers for was the comic section and other than that the papers were full of politics which was confusing. The exam was not only a test of knowledge but also of speed of thought. It was enough to daunt any nineteen-year-old college kid in his final year of college, but Jay had a new steely resolve.

It seemed like just the other day that his father had come back from office and asked Jay's mother where he was. Jay had just got back from college and was eating lunch.

"Look at that useless loafer just sitting and eating free food..." said Colonel Shankar, in a familiar opening gambit.

Jay's mother looked on nervously, as mothers do when they have a son and a husband both approximately the same size behaving like bulls. Choosing between the bull you married and the one you gave birth to become difficult, but Indian culture trains you for just such an eventuality. And thus it was that Mrs. Shankar stood by with a faux demure look, as her husband continued.

"How much longer are you going to just laze around like a useless rowdy (Shankar senior's favourite and ultimate adjective)? When I was your age I was already in the bloody NDA!"

Jay was now thoroughly incensed, "Maybe your parents didn't want you around the house...Don't worry I am leaving as soon as I can. Who wants to stay here in this dump?"

Colonel Shankar had to have the last word. He turned to his wife and said, "Tell this useless bloody fellow he will not get any free food after graduation.

We should have stopped feeding him after he turned eighteen, but next year..." And thus the die was cast.

Jay's feelings and ego were hurt and he was now determined to prove himself. He went to his neighbour Jas's place, a local legend in the cantonment. This worthy had appeared in the NDA exam at least three times and the IMA exam once but flunked in the SSB every time. Jas had finally given up and appeared for MBA and was joining some IIM (Indian Institute of Management) next month.

Jas handed over all his study material and notes along with his best wishes which were welcome. He also unfortunately had a lot of free advice to offer which was not. The fact that this free advice was coupled with an aggressive personality and intense disappointment at not having got into the army...did not make the interaction pleasant. But Jas enjoyed Colonel and Mrs. Shankar's confidence and thus had to be borne.

Jas swaggered into Jay's room and plonked his ass down authoritatively. Jay waited patiently for what was to come.

"IMA is ok but you know why the chicks really dig the MBA types...and you know how many babes there are in the IIM?" asked Jas with a smug look.

Jay did not really know the answer nor did he want to know, but he had a feeling Jas would tell him anyway. Again...For the twentieth time. And he was right. Jas started off, "MBA guys really earn the big bucks, and you know man my starting pay will be...."

After one hour of MBA grad emolument package dissection Jay could not bear it anymore and said, "Hey man, I got to go take a crap..." That had the desired effect and Jas left. He looked a bit disappointed that he hadn't impressed his audience more with the financial package of a potential MBA graduate. Jay went inside the bathroom and revelled in the peace as he let his sphincter and imagination loose.

He thought about all the fun the young officers, especially the bachelors, had with girlfriends galore and money to burn. Dire Straits, his favourite band, had said it all in their song 'Money for nothing and the chicks for free'. Jay would later learn that the Dire Straits were talking about rock stars not army officers. But as far as Jay was concerned army officers *were* rock stars.

He thought about the way they swaggered around in their cool uniforms, got to fire machine guns, had all those nude posters pinned up on their walls. That was the life, party all the time and sometimes in those gorgeous mess dresses. He was picturing himself in a 'blue patrol'[2] when he heard his father knocking at the bathroom door.

"Stop wasting time and get back to your preparations!" said Shankar senior, then added, "And get ready, some people are coming over for dinner."

Jay wondered at his father's concept of time and its gainful use. Essential tasks like eating and crapping came under the category of 'waste of time'. But evidently being polite to a house guest was not a waste of bloody time, and he seemed to be doing a lot of that.

The thought of visitors made Jay groan inwardly. He of course would be expected to get dressed, stay in his room, study and only emerge when dinner was served. That was usually the worst part, having to make polite conversation. He really missed the time when he was younger and expected only to be seen occasionally, and never heard or have to hear.

The guests, all military men, would either tell him to 'buck up' or would start some inane story about how fun life was in IMA. The women would either tell his mother how hard their sons worked for the NDA or IMA or some similar story about their successful sons or daughters. Either way he had to maintain a constipated yet dutiful grin on his face till it ached or he lost his appetite.

A month later, he felt somewhat the same, when he entered the exam hall in Chandigarh. He had never seen so many turbans and beards together or separately in his life. The exam that day was English, but the volume of Punjabi being bantered about could have fooled anyone. He looked at the guys in front of him; they were mature-looking to put it mildly. He was aware that people in these parts generally had no concept of age and sent their kids to school when they felt that the child was old enough to walk to school on its own.

The last age for joining the IMA was 22, but most of these guys looked much older. Jay himself was but 19, and would turn 20 in a few months, by which time he hoped to have graduated. He mentally shrugged, maybe it was all the facial foliage which made these guys look older.

The other thing that struck him odd was the seating which was supposedly random and based on a computer distribution, and yet.... all the guys sitting in the front and at the back as well as in the other rows seemed to know each other really well. He and a few other *fauji* kids seemed to be the only strangers. The army was probably more of a way of life in these parts thought Jay to himself. It felt like entire villages had turned up for the exam.

Jay asked one of the men sitting on his seat to kindly get up. That chap looked positively offended and asked, "*Kya jaldi hai, oye*?"

Since the announcement to take one's seats had already been made and the guy was about six feet tall, Jay supposed the question was rhetorical. He backed off and waited patiently while the group around his seat conquered English grammar through Punjabi dialogue.

The exam was really predictable and Jay mentally thanked Jas for all his tips. After having gone through thousands of questions in the CDS guides and competition magazines, there were no surprises. Jay wondered if the lack of imagination in the people who set the question papers was a precursor to what he would be facing in the army. He was partially right...

That evening he compared notes with three more fauji kids who had taken the exam and were staying in the command mess. They met in the TV room and after a round of introductions, the ice was broken by a chubby lad called Akarsh Ghai. The other three were from the same cantonment of Pathankot and knew each other well.

Master Ghai was all for going and grabbing a few drinks at the bar. The others, Abhay Lakra, Shiju Abraham and Jay were hesitant and apprehensive, partly because of the math test the next day. Also the possibility of being discovered and booted out of the bar was very real since command messes have very few friendly or young officers. The personal embarrassment was an inconsequential factor; one develops a thick skin living under a military father. However retribution once the same military father learnt of the escapade and was himself embarrassed would be substantial.

"*Chalo* (c'mon) let's go and get a drink," said bold Ghai. He was relaxed, having made up his mind to go only for OTA. He only had to pass English and general knowledge, and thus not too at all worried about the math part of the test. He planned to mark all the multiple choice answers 'b' and leave the rest to mathematical probability.

"Forget it *yaar. Kal hamara* math test *hai* and *yahan pe gaand phati hui hai*![3]" said Lakra.

Abraham looked equally worried, "My pop told me this was the last chance and after this he was going to send me to the regimental centre and get me recruited as a sepoy."

The others looked at him with horror. Jay knew what Abraham was talking about. Col. Shankar had once taken Jay on a tour of the regimental centre. The sight of the recruits constantly on the move, harried by instructors with thick canes had scared the shit out of him.

"We gotta study man, but how do you plan on getting a drink? You think the barman is going to give you a drink straight away?!" asked Jay though he knew the answer.

Pat came the answer from Ghai, "You fuckers are totally not officer material. Note my dress and my moustache carefully, and tell me do I look like a young captain or some sorry fauji kid here to write the CDS exam?" With that captain-designate Ghai lit up a cigarette.

The others looked at Ghai with new-found respect as they noted his sparse albeit carefully trimmed whiskers, 'dark pant and light shirt', that unstated dress code for the mess in the evening. Ghai sahib had evidently planned well ahead, also the sign of a good though degenerate officer. But there were still a few lingering doubts in the minds of the apprehensive yet thirsty army aspirants...

"Your frigging dress is ok, but what about your shoes, you silly bugger?" asked Lakra.

They all looked down and lo behold Ghai was wearing a pair of black Addidas, albeit highly polished, maybe it would pass muster. Ghai shrugged it off. "Next question?"

"And what about the bar chit...don't you think the barman is going to wonder when you sign it as Colonel Ghai, your pop's name! Anyway this is the command mess, no young officers around..."

Ghai replied, "Watch and learn you buggers. If the barman was so smart and educated he would have become an officer by now, and we are in a command mess where the only young officers are ADCs (aide de camp),ha-ha!" and with that chuckle Ghai vanished into the bar. He appeared a little later, accompanied by a waiter bearing rum and coke for everyone. The old waiter looked a bit suspicious but smiled indulgently at them and laid down their drinks.

"Drink up guys! It'll help you reeeelax!" drawled Ghai. "I also told them that I was ADC to XI Corp Commander and you were his nephews, and now complimentary snacks follow..." The others looked at Ghai with new-found reverence. The kid was an artist and had a bright future in the army. Any man who could look after his peers would definitely be up to the onerous task of looking after his superiors. They did not realise that the barman probably knew all the ADCs well. Luckily he was sympathetic to their cause and more positive about their future prospects in the army than they themselves were.

II: Free Advice

Jay was relieved that his final exams were over. The last few months had been hectic. He had gotten a rude shock when he had been informed that his attendance was not enough for him to sit for the final year exams. Luckily he had been sending bottles of rum to the clerk at the college office to cater for just such an eventuality.

After all there was precious little that they taught in college that was going to help him in real life. The clerk did not need a college degree to know that attendance was more important that actual college learning. Also, his conscience had been more agreeable when he learnt Jay was skipping college classes for a higher purpose...to join the Indian Army. The rum did not hurt either.

Jay had joined mid-session when his father had been posted there. There had been the initial running around to get the inter-university transfer certificate, no-objection certificates, good conduct certificate etc, etc, all those pieces of paper that petty bureaucracies invented to complicate life. But Jay had managed and that had been possible thanks to a liberal use of Old Monk rum to lubricate the process. But all that would come to naught if he did not get a degree.

The attendance problem was sorted out but there was still the delicate matter of Jay getting a 'high' (as Jay tried to put it to his mom) III division. A III division degree would not look good in any interview except for the post of cowherd; at least that's how Col. Shankar put it. Jay was made aware that the only thing a III class degree was good for was for the service industry, cook, waiter, cleaner etc. This was a life or death situation; at least for Jay. Col. Shankar stepped into the breach and squared things with the principal. The principal had thawed after Jay's father informed him that Jay had passed the CDS exam and was just about to become an army officer. He conveniently forgot to mention the fact that Jay was yet to qualify in the SSB. Well...one thing at a time.

The principal gave him a rousing talk on patriotism and also the grace marks required to get a II division. Everyone was happy. Everybody had been real nice to Jay since he had passed his CDS exam. He really, really wished this phase would continue. It would... if he got through the interview.

Col. Shankar had arranged for one of his course mates posted in the station to coach him for the SSB. Lt.Col. Bhargav was a crusty old-timer. He had missed his next rank but was still highly regarded among his peers. Many also regarded him a constipated ass. Bhargav uncle was still a bit sore about the ridiculous number of 'nincompoops' who had risen higher than him. He made this very clear to Jay at the very first meeting. He also made it clear that Jay was now more or less in the army and thus no longer a kid, specifically a senior officer's son.

"I'm not 'uncle' anymore, so kindly get used to the idea of calling senior officers, 'sir'."

This was Col Bhargav's opening sally as Jay stepped into his office. Col. Shankar had warned his son to be confident at all times since Bhargav wanted to create the interview atmosphere.

There followed a long silence while Jay stood fidgeting nervously in front of Bhargav's desk. Bhargav looked at him intently for about a minute which was 55 too many uncomfortable seconds more than required. It was evidently part of the interview coaching.

"Stand straight, don't slouch like a loafer! When the interviewing officers sees you they should see someone who looks like a smart officer or a man he would like to see in his mess."

Jay straightened his shoulders threw out his chest and stammered, "Yessir".

"Louder! Show some spirit and why haven't you wished me? Now get out and let's start all over again."

Jay felt like telling the old man to fuck off, but that was not an option. So he went back in, knocked at the door and asked in as gruff a voice as he could muster, if he could enter. He then entered and wished Bhargav brusquely just like he had seen the Germans do in the World War II movies. It worked, Bhargav beamed and said, "Good, Good. Exactly the right tone one must not appear too servile or 'Bolshie[4]'!"

Jay felt better and made a mental note to keep a brusque Teutonic tone while addressing the interviewing officer. The next piece of advice from Bhargav was, "Kindly smile a bit and don't look so bloody constipated."

This was a bit hard for Jay to take from Bhargav who had not made it to the next rank thanks to constipation in looks and attitude. At least that's what Jay had overheard his mother tell his father. Clicking heels, small bows and slightly servile looks or smiles were part of officer-like deportment it seemed.

"What are your bloody interests...do you have any?!" was the next question.

Jay wondered how much his father had actually discussed Jay' shortcomings with this old fogey...whatever it was, it couldn't have been too good.

In an attempt to sound original and not a complete poodle-faker[5] as his father had no doubt portrayed him, Jay replied, "Uhh... horse riding, squash and tennis..." He hoped the answer would be classy enough to consider him ideal officer material.

Bhargav gave him an exasperated look and confirmed an opinion Jay had dreaded, "You sound like a real poodle-faker!"

"Look son..." Bhargav said, "They are looking for leaders of men and young officers, not sissy-faced generals!"

There was another uncomfortable pause and Jay heard a phrase he would continue to hear for the next few years very often, "Pull up your bloody socks."

"Tell me any worthwhile interests you have, any 'troop' games...games you can play with the men. Be honest you bloody joker[6]!"

Jay thought carefully and replied, "Yessir—Basketball and toy making."

Jay had played basketball a couple of times because some girls, three really gorgeous sisters used to play; actually they just came to watch their brother play. His toy making credentials were based on the one attempt to make a gun when he was in kindergarten. Jay's mother was still proud of that attempt and preserved the relic for posterity, and Jay's enduring embarrassment. But he could now say it honestly.

"What's the size of the hoop?" and after a pregnant pause Bhargav sir added, "It's found commonly in the basketball court..."

Jay looked down and admitted defeat. He knew the basic basketball rules, but wasn't referee material. He felt a bit peeved that the whole point about games being played for fun was being missed. 'The only numbers that should matter in any sport is the score not the bloody dimensions of the field', thought Jay, but wisely kept his thoughts to himself.

"You said toy making... now kindly make me a toy, you got five minutes." And Bhargav pushed over a sheet of paper and some paper clips towards Jay with a triumphant look. Actually it wasn't really triumphant, but more the kind of look a cat might give a mouse which it has finally cornered.

The look was a mistake and now the cornered mouse-Jay—was riled and inspired. He actually managed to make a decent enough soldier complete with a paper clip rifle and bayonet.

"Soldier in snow camouflage..." said Jay as he pushed it towards Bhargav.

Bhargav was mildly impressed and thawed enough to give Jay some more constructive tips. The tipping lasted through the afternoon as Colonel Bhargav got carried away by his noble task of preparing young Jay for his IMA debut. The carrying away was interrupted by thirst and Jay was forced to accompany his new-found mentor to the mess for further mentoring.

Jay knew what was coming and tried his best to avoid it with thought provoking excuses. Bhargav would have none of it—he had not had such a captive audience in years since most of the junior lot avoided him in the mess. Jay tried but ultimately realised that he had very little say in the matter, literally. Bhargav would not let him get a word in edgeways. Jay wondered if army interviews were actually monologues.

They reached the bar where a few other old burnt-out fogies were downing gimlets. They hailed Bhargav and looked at each other meaningfully. They recognized the fact that some new meat had been brought for sport. Jay also recognized the look, he had seen it in Bollywood movies when the villains eye the captured heroine with no hero in sight.

Jay spent a long hungry afternoon cringing in one corner of the bar. He was rationed only one bottle of beer to slake his ire and thirst, as he bore the sustained assaults on his ego and military prospects. He heard about the importance of the interview in the SSB. How the interviewing officer was more important than the GTO or group testing officer. The reason was that the interviewing officer was usually the one who wrote the GTO's ACR[7].

Jay could see the importance of the logic but not how this piece of logical gossip was going to help him get through the interview. The afternoon wore on and it was at the very end that he got the only significant piece of advice of the day.

"Listen young man," said the old colonel, "these interviewers have seen thousands of young men...you can't fool 'em. Just be honest and try and show 'em some potential if not outright OLQ[8]!"

Another oldie took over from there; it was really like a mental gang rape thought young Jay. Oldie number two said, "Remember... they are looking for social ability, self-confidence, and cooperation and...about fifteen other qualities. I guess some of us didn't have it. Ha-ha!" The last ha-ha had a bitter note.

The assorted oldies there guffawed and one of them added, "Old boy, the SSB is about getting good young officers, not frigging generals!"

On that note, what felt like a brainwash with gin, came to an end for Jay. The only thing he could remember were gin fumes, boring old men and a vague apprehension that life in the army as a young captain may not be all adventure. Back home in the evening he looked for respite but was disappointed. His parents were waiting with a determined look. That was the look they always had after a vigorous discussion of his future prospects or present day transgressions.

These discussions were usually early in the morning or sometimes when they thought Jay couldn't hear. The late afternoon sessions usually did not bode well. It meant dad had not been able to have the ritual siesta so precious to the military man who is usually short of sleep. It also meant mom had forgone her afternoon soap opera or mahjong session to reach some important decision on her progeny's welfare. The decisions usually upset Jay's lifestyle and tranquillity such as it were. "You're fat!" opened daddy Shankar.

Jay looked at mummy Shankar pleadingly. The sad puppy look had almost always worked in the past. Like most mothers anywhere in the world Mrs. Shankar was never sure if she had done her children's appetites justice. This was particularly true of Indian mothers who had produced their sons after countless hours of labour, at least more than other women in the civilised world.

In the past any reference to Jay's slight paunch would have brought Mrs. Shankar's combative spirit to the fore and she could have been counted on to make a snappy reply on the lines of, " The men in my family have always been big" or at least, " It's just puppy fat". But it was not to be this time.

Actually, truth be told Jay had become a big puppy and after the last few months of studying, and the subsequent pampering after getting through the exam had increased his girth noticeably. Col Shankar had noticed as had the doctor at the military hospital. Jay had been taken for a pre SSB medical check-up to rule out any rude shocks leading to medical rejection after getting through.

"Aww...come on dad how overweight, just four kilos! I don't think they'll take it that seriously...It'll get covered in the ten percent allowance."

"If you get rejected for weight, you'll be given time to lose weight and go back for the medical. BUT IT'S ONLY THE STUPID WHO TAKE THAT CHANCE!" said Col. Shankar, a bit shriller than he had intended.

It was decided and Jay was given the choice between a civil life and civil disobedience. There was only one week left for the interview. Jay spent that time wisely, practising public speaking in the bathroom and before every mirror in the house. The remainder of his time he spent reading books and magazines that would improve his knowledge. He hadn't read a *Vogue* or *Debonair* magazine in a while; it was only newspapers or stuff like *India Today*.

The gaps in this dark phase in his life were filled by salad, actually lots of salad. Mummy had drowned her guilt by ensuring young Jay never went hungry. But if there was a fate worse than hunger it was salad burps. Jay imagined his burps were actually beginning to sound like bleats or buffalo bellows. He dreamt only of uniforms, bikes, guns and food and not necessarily in that order.

NOTES

1. OTA (The Officers Training Academy) in Chennai, Officers on commissioning were bound for a five year service stint.
2. Blue Patrol is the name of the winter mess dress for officers of the Indian Army. It gets its name from the dark blue dress duty officers wore in the late 1800s to the mid 1900s to check the guards and patrols.
3. The exact translation would be 'here our ass is getting ripped', usually a term to indicate fear, nervousness or apprehension.
4. 'Bolshie' or short for 'Bolshevik' a common term for someone who is a bit of an upstart, Slang handed over by the British used by a senior generation officers, not so common now.
5. This was old fashioned slang which as per the Harper-Collins dictionary meant-'a young man or newly commissioned officer who makes a point of socializing with women; ladies' man'. Probably an apt description, in this case for young Jay.
6. Another favourite derogatory or and colloquial term used to refer to juniors not really held in high esteem by officers of the Indian Army of a certain vintage.
7. ACR (annual confidential report), the basis for further career progression.
8. OLQ (officer like qualities), for example—honesty, drive, determination, initiative etc etc.

Chapter 3

THE FIRST STEPS

Jay sat in the second class compartment of a train. This was the first time he was travelling alone and he was a bit uncomfortable, especially after the long list of do's and don'ts handed out by his parents. If he'd actually listened to even ten percent of the points on that list, he would have most likely been arrested for wilfully impersonating a psychotic individual.

His father had always adopted the bluff 'don't mollycoddle the young man' attitude. This was now replaced with a 'follow my briefing and you will stay alive' type act. Actually the old man did not say much as he was feeling proud of his son for a change. There was hope he might become a decent officer and maybe even a gentleman.

His mother didn't give a damn about all that; she was apprehensive about sending her only son alone. She had insisted Col. Shankar send a man on temporary duty to escort Jay to Allahabad. Thankfully the old man had resisted. Jay shuddered at the thought of getting down at the station from a second class compartment with a chaperon. What would the other candidates think...definitely not good things?

The guy next to him seemed to be '*fauji baccha*'[1] type. He was also wearing smart officer-like clothing aka light shirt, dark pants and OP (Oxford pattern) shoes. It seemed a bit incongruous in the current gathering of sweaty farmers and unwashed labourers. But what the hell, it was in situations like this that the true officer-like spirit shone. '*Fauji* baccha' next to him demonstrated said spirit by squeezing out a few more inches of the seat.

It would be pertinent to mention here that the seat was designed by the

Indian Railways to comfortably seat at least four healthy Indian bottoms. But the population explosion meant defying the railway coach designer's paradigms. There were six people already sitting on the seat. It was thus a very significant gesture, those few inches squeezed out to make a fellow would-be soldier comfortable. Jay acknowledged this gesture by introducing himself and thanking his 'officer like' benefactor.

His name was Rakesh and he was a 'Texan[2]' and he was not a *fauji* baccha as earlier thought. Jay had been fooled by the dress and eager air about him. Rakesh was a true son of Haryana and desperate to get into the army. His father had been unable to join the army, unlike the other male members of the family of the last five generations. And now it was up to him to avenge the blot on the family's martial history. He had already avenged it in part by being a national champion in Taekwondo.

He kept looking around furtively and had a generally pained expression. Jay finally asked him if he needed to go to the bathroom. Jay thought Rakesh looked like he had an upset stomach. But the reply almost upset Jay's stomach.

"You gotta be on your toes all the time '*chorei*'[3]...these SSB chaps have people everywhere watching all the candidates. You have to show OLQ type points all the time", Said Rakesh. The way he said OLQ as delivered in the Texan accent sounded more like some rural laxative.

Jay was not sure he followed and wasn't quite sure if the Indian army would go to such desperate lengths to get the right candidate. But still he didn't want to annoy a psycho taekwondo Texan. The rest of the journey he humoured Rakesh by agreeing with him. His neck was not designed for so much nodding, so he pretended to doze off.

It seemed the closer he got to the army the more crazies he came across. It turned out this was Rajesh's fourth attempt at clearing the SSB. Jay silently acknowledged his superior experience and Rakesh, also silently, assumed the role of Jay's mentor for the next one week

The train reached Allahabad station and Jay felt butterflies in his stomach. Rakesh looked grim; he had recognised a few other eager competitive faces. Jay realized that there must be many more experienced hopefuls like Rakesh on that train. Jay took a deep breath and dragged his bag out.

There was the usual mad scramble to get to the door even though the train was slated to halt for half an hour at the station. Jay wondered if this was part of the 'OLQ' requirement or just the typical Indian mentality thing. He sincerely hoped somebody was watching him showing initiative and forethought at joining the mad struggle to be the first to get off the train.

They had instructions to report to the MCO[4] office at the railway station. They found a couple of bright-eyed NCOs[5] along with a JCO manning a table outside the MCO office. There was a blackboard which said boldly 'Welcome SSB Candidates'. The look on the reception committee's face was anything but welcoming. They looked at the new candidates up and down with undisguised scorn and authority.

The reception committee was undoubtedly savouring this moment of power. The assembled gaggle of candidates meekly got into line with a fluidity that amazed Jay. The onlookers at the railway station were also watching in amazement at the disciplined behaviour like queuing up, as yet unknown in India.

Jay stood gawking for a few seconds before he remembered what Rakesh had said on the train about being under observation all the time. It had worked in other dictatorships and evidently the myth was working well here too. And Rakesh was right there prodding him in the kidneys to remind him.

The initial paperwork started. They showed their call letters and signed their names in the nominal rolls. They were then directed to the army buses standing outside. Some of the candidates seemed hesitant. Rakesh evidently knew what was coming, because he took Jay, who was still gawking at the scene, by the elbow and hustled him to the bus.

"What was that about?" asked Jay.

"Same thing every time...a bunch of *chutiyas* think they have already become officers just because they have got through the exam. And now they want someone to carry their luggage to the bus!" replied Rakesh a bit smugly.

Right enough they heard a loud shout followed by some indecipherable haranguing. The tone was clear even if the words were not, and a few minutes later the hesitant group was climbing into the bus with their luggage looking sheepish in the extreme. The others looked at them with a mixture of pity and or sympathy. Everyone knew this flock was doomed.

One of them, a smart looking kid took the seat across the aisle from Jay. He stuck his hand out and introduced himself as Rohit Kapoor. He seemed like a cheerful sort, even if a bit confused. Turned out that master Kapoor was also an army officer's kid. Some of the others, who had heard the introduction, gave him hostile looks.

The hostile lookers seemed to be the Chandigarh crowd and some of them even looked familiar. They were having a heated discussion in Punjabi about the unfair advantages '*fauji bacche*' had in the SSB. It was an old complaint and not always true but looked like even Rakesh believed it. They seemed eager for a fight and were aggressive as hell. A fight would automatically disqualify the participants and the Chandigarh chaps well knew this, but were slave to their genes and mob machismo.

Jay was getting the impression that there were primarily four groups that came for the SSB. The fauji kids were one obvious lot; there were the Jats and the Punjabis who were professional SSB candidates. Then there was the lot from the Sainik and military schools[6]. These two lots had been groomed and/ or brainwashed from an early age to revere soldiery as the ultimate profession. They were deadly keen, almost desperate to get into the army.

Then there were the professional 'competition masters'. This distinguished group had appeared in almost every competitive exam in the country. They were here either because they had got through this exam first and wanted to try their luck, or had not got through any of the other exams like UPSC, IIT JEE, CAT etc etc. They were not here because they wanted to serve the country, no sir! They were here because their preferred method of making it big or at least making big money in the IAS or IPS had not worked out. The only other honourable way of earning a living as a class I gazetted officer[7] was as an army officer. Too bad they would not be getting that fat dowry or the promised fat off the land. No wonder this crowd looked a bit disappointed.

"Hey man," he heard Kapoor say, "haven't done any prep for this interview, my old man has been on my case all year..." and he kept up the chatter till they reached their destination.

They reached the SSB and were unloaded with what seemed like excessive martial noises and other such barking sounds from the NCOs. Even the civilian sweepers and peons had joined the fun to gawk authoritatively at the new meat.

After some more paperwork, mostly submission of certificates, they were told which barracks they were allotted and which group they were assigned to. They were told to check up at the notice board for their chest numbers. It turned out Rakesh was part of his group. It was weird the way guys were going out of their way to be polite to each other, even kind. Jay found Rohit standing near the bus looking dejected and walked up to him out of '*fauji baccha*' solidarity.

"Decided to drop out?" he asked.

"No way man... you kidding? My old man would kill me if he even thought I wasn't serious about joining the Army. As it is he's gonna be mad I forgot my tenth standard certificate." Replied Kapoor dejectedly.

Jay too shook his head sadly. This was serious indeed. The tenth class certificate was the ultimate proof of age for the Army, indeed most government jobs in India, to get a passport maybe even to prove paternity.

"So what now?!" He felt a bit sorry for Kapoor.

"Well they've given me a different date and told me to come back with the certificate".

"Good luck man," said Jay.

"See you in IMA" Kapoor replied and they shook hands. Jay had the feeling they would meet again.

NOTES

1. 'Baccha'—kid or in this case denote 'army brat'.
2. Texan—Jats, a martial community of North India and generally the people of Haryana are called Texans, probably because of their 'lean and mean' looks and reputation.
3. Chorei—Texan slang equivalent of 'kid' or 'boy'.
4. MCO (movement control organization), an office of which can be found in all railway stations with an army base nearby.
5. NCO/JCO (non-commissioned officers) like corporal, sergeant etc and junior commissioned officers above NCOs but junior to commissioned officers.
6. Each state has Sainik schools run like any other military academy. There are also five military schools earlier known a King Georges military school and RIMC—Rashtriya (national - previously the 'R' stood for royal) Indian Military College. All have produced distinguished alumni who have risen to high ranks in the Indian military.
7. Class I Gazetted Officer, denotes the highest class of government officers. All the most sought after central government services like IAS and IPS are on par with military officers, being class I. The difference in perks and career progression is of course phenomenally different and thus the immense popularity of central government services over the Army.

Chapter 4

SSB OR SCREW AND SEND BACK

The documentation seemed to carry on forever. Jay didn't know it but he was being put through the first management technique of the army, 'hurry up and wait'. The candidates were addressed by the group testing officer or GTO on duty, as he introduced himself. He had a whole list of do's and don'ts, mostly about dress and deportment.

Jay felt a bit nervous when he saw some of the others actually taking notes while the duty GTO spoke. They displayed a genuine eagerness that made Jay feel guilty and inadequate. Yet again, he reverently hoped that he did not look like a poodle faker. He had taken an extra-short haircut just to a ward off such an impression.

They spent the rest of the day doing pseudo psychological tests and had a group discussion. This was all part of the initial screening. His group had this really pushy guy who Jay later learnt was a soldier trying to get into the ACC[1] Wing of IMA. The guy totally lost the thread of the discussion and kept trying to sell something he had probably memorised before coming to Allahabad. It was pathetic; somewhat like having a tree as the topic, talking about a cow instead and saying it's tied to a tree to connect the two.

It was tough but Jay managed to get in a few words in between when he thought he had something logical to offer. The next day was another round of psychological testing. They were given a whole lot of words and asked to make sentences, then pictures and then a story. The biggest hassle was to make it all sound positive. Although he had had some practice, the time available was not enough to either remember any story mugged up earlier or make anything up afresh. You just had to write what came to your mind and that was what they wanted. It's not easy trying to write something positive about a picture with a corpse.

These mind games carried on till lunch time, followed by some more paperwork where they were asked to fill up lengthy forms. These were basically to get as many personal details as possible to help the interviewing officer. This last piece of information was from Rakesh who urged him to write really 'good details' about himself. Jay had been warned about this by Bhargav. He would have to corroborate whatever bullshit he wrote for the interviewing officer, who would be able to spot the pretensions. But still Jay could not resist adding that he had been captain of the college basketball team.

That night Jay felt nervous as he watched the 'Chandigarh Boyz' like as they liked to call themselves as though they were some Chicano gang. The others referred to them as 'Minerva[2] *de munde*' (literally 'Minerva's brats' in Punjabi). Watching them discuss the next day's events among themselves was a revelation of sorts. They were all secretive and huddled as if discussing an attack. It was obvious from the way they shot venomous looks at him that they considered Jay a poodle faking *fauji baccha*. Jay would have been relieved to learn that it was nothing personal; the Chandigarh 'boyz' considered all non-Punjabis as 'English speaking sissy types', and army officers' kids especially came in that category.

They were heavily under the influence of Punjabi rock or rap. To keep in sync with that genre they had cut their hair real close around the temples but left it long on the top. That may have been a good idea when you're covering it with a military cap. But when you have the South Asian tendency to get chubby faced it makes one end up looking like a butter nourished mushroom. But they did not care, the SSB coaching academy had told them 'appearances are of minimum interest'. This was only partially true.

The next day was the group testing which would carry on for the next three days. The groups and chest numbers had been put up on the notice board in the mess. Jay felt he was already in the army. Just two days and he had been reduced to a nameless number in a shapeless dress already. He felt a nice warm feeling creeping up his spine.

Jay went to check out his games dress, specially stitched for the occasion. All white of a pre Second World War pattern, tailored half sleeve shirt, half pants, and that meant half pants not those pansy shorts most youngsters wore these days. The shorts could have doubled as a divided skirt anytime. Shape-wise they could make any RSS *pracharak's* khaki shorts wilt in shame. Lastly there were white sports shoes and socks.

There had been words between Jay and senior Shankar regarding the half pants. Jay had finally managed to reduce the length by a few inches above the knees. Though it still looked like something out of a Hollywood musical and got drafty even with just a ceiling fan at slow speed.

He had not wanted to look like 'Charlie Chaplin goes to PT', while daddy was convinced that it was the only way to look officer like. Jay begged to differ, but had to choke down harsh phrases like 'stuck in a time warp', and 'out of touch with reality'. He had needed goodwill and pocket money, not a fight at this point.

And so the old man had had the final word and managed to get an all-white pair of PT shoes, without any symbol, logo or design. It was with great difficulty that they'd found it, and Jay silently swore that he would personally find some museum they could donate it to...after he had finished his SSB. But he had been unable to refute what his father repeated ad nauseum, "I have got through SSB in my first attempt and I know what I am talking about."

He went to his dormitory that was in a pre-independence building with a depressing dearth of lighting and surfeit of mosquitoes. The kid next to him was almost in tears. His name was Chib and he was from Himachal Pradesh. He had been told by his neighbour, a huge burly Sikh, that he would be disqualified since his shorts had a racy stripe down its side.

Jay tried to console him, "Hey c'mon, they will fix you and your shorts when you get to IMA. They just want to make sure we all wear the same stuff generally and not seduce the selection committee with sexy shorts and stuff!" That was actually the truth but Chib did not seem convinced. His face and attitude conveyed as much.

"That inspector sahib has been to many SSBs before. I'm sure he knows what he's talking about," sniffed Chib.

Clearly Chib had chosen a mentor already. Jay observed the *Khalsa* [3] who, as Chib informed, was Inspector Malkeet Singh of Punjab Police. Inspector sahib was a former sports man in Punjab and was here...just to give the Army a sporting chance. He had in fact given the Army several chances to employ him and save the land. Seeing him snore with his massive paunch and beard quivering, it looked as though he had been living pretty well, off the land and off the police. Jay shuddered; there was no way inspector *sahib* was below twenty five. Yet here he was an old wolf in an ill-fitting lamb's clothing.

Jay was sure that even Sardar Malkeet's mum was not sure about his vintage. She would most likely date the momentous occasion of said Sardar's birth with a more significant landmark like maybe a hurricane or an important calving. It may not sound as important to the average Indian, but these were life and death issues for the rural Indian.

Going by inspector Malkeet Singhji's prodigious girth, it could not all have been mummy*ji's* contribution. Likely some blessed cow that had shared its childhood and milk. Jay had an awful premonition and he rushed to the notice board to check if the inspector was part of his group. He was, and so were Chib and Rakesh. Jay was less than elated.

The next day was a mad rush as Jay had to finish what used to be a leisurely morning routine...in one-fourth the time. This was thanks to the bathroom being commandeered by the professionals and the 'boarding school types' who were used to unearthly times like 5.30 A.M. for getting assembled. It was chaos as pampered sons and unruly college students tried to shave, while standing in line for a crap or a bath. Some made it for breakfast most did not. They learnt another military lesson, 'being clean isn't important... being on time is.'

Jay and his group of six got together under a tree at a stand. Theirs was a very diverse crew. Jay keenly observed the others in what he hoped was a discreet manner. The others did the same with varying degrees of discreetness. Then they all simultaneously remembered that they were supposed to display friendliness, positive cooperative spirit and of course, sociability. These were all part of the qualities that the SSB was hoping to discover in the candidates, and it was possible the GTO was watching from somewhere. There was a scramble of introductions.

Apart from inspector Malkeet, Chib and Rakesh, there were two more potential wannabes. One was a brainy, bespectacled and paunchy kid from Andhra called Ramulu. Ramulu was also unused to early mornings and was actually unshaven. The others looked horrified at this blatant display of 'unofficerlike' deportment.

It turned out that Ramulu was an engineer and had appeared for almost all the competitive exams of the land and was waiting for the results. The CDS people were however more crafty and declared their results earlier thus luring people like Ramulu to try their luck. That is the temptation of getting a job in hand with its assured stability, and thus unwilling to take a chance with other selection processes. It was actually a pretty smart and tricky move by the military, second only to the press gangs[4] idea.

The sixth man, so to speak, was Dilip Singh who had a decidedly military air about him. He was actually from the corps of EME[5]. The man had got through for ACC but since he had already done his graduation, had decided to save time and join through the IMA. He looked supremely fit and confident. Rakesh and he faced off as potential rivals. Both had prior SSB experience and would now be fighting it out in the group discussions. Jay realized that in this testosterone radiation there was bound to be collateral damage. He just hoped it would not be him.

The inspector called the group to order and suggested they should all cooperate and so on. Basically what the good inspector was trying to suggest in his inimitable way was it was going to be 'his way or the highway'. The others did not look too happy but nobody wanted to argue with someone with so much brawn and aggression. The GTO came over and introduced himself and asked the group to do the same. The air was thick with tension. Jay found the whole thing exciting except when his father's face popped up in his mind from time to time exhorting him to pass or... Jay was wondering about plan 'B' when he heard the GTO addressing him. He nervously got up and introduced himself. As he did so, he could feel the hostility from the others when they heard his father was in the Army. Why? Because they all took it for granted that he would get through using 'pull'. But the glint in the GTO's eyes told Jay and the others this was going to be a tough ride...for everyone.

They started off with a group discussion. The topic was 'India's nuclear policy'. It was fairly innocuous, but inspector Malkeet started off the dick-waving contest with some really dumb statement. Most of his pronouncements seemed to have an unspoken 'or else' threat. Even this latest inane one that, "India's nuclear policy is the good policy!" He topped it off by glaring around at the rest of the group as if daring anyone else to disagree with him.

Things went downhill from there as far as group dynamics was concerned. Jay managed to contribute a few sentences and surprisingly so did Ramulu who actually stood up to the inspector.

They then moved on to group tasks. That was when the shit hit the fan as far as the group was concerned, the inspector himself was unconcerned. The group had been given a gigantic 'anaconda' made of gunny sacks and asked to take it across a set of obstacles as a group. The idea was to observe group dynamics as per the Bhargav testament.

It did not work out as planned at all. There were at least three guys with a plan. They all got it wrong. Jay, Chib and Ramulu were all prepared to accept whatever sensible idea was given and get the thing moving. The remaining three, inspector Malkeet, Rakesh and Dilip were convinced that they had to show only leadership traits here by making the others listen to their commands.

The clash began to take on regional undertones with Rakesh and Malkeet facing off in a replay of the traditional rivalry between Punjab and Haryana. The GTO was beginning to fidget and Dilip joined the other three and told them to start off. Rakesh also joined in but the inspector continued to bark orders.

The crowning glory was when they were negotiating a particularly difficult obstacle involving a high wall. They had managed to get the 'snake' over with some difficulty. Jay was still on top of the wall when he felt someone tugging at his ankles. He looked down to see a fat red face with a frazzled beard. It was the honourable and out of shape inspector Malkeet Singh who was now beginning to feel the wear and tear caused by too many 'Patiala pegs' and butter chickens.

"*Abhey mujhe bhi khincho behanchodon,*" (pull me also up you sisterfu...) wheezed the inspector.

The fact that he had used a reasonably mild expletive '*behanchod*' indicated inspector sahib's mind was as frazzled as his beard. The use of bad language and that too while addressing other candidates was a strict 'no no'. The GTO was hovering nearby but the inspector was beyond caring. The heat, dust and exertion had been too much for him. Jay tried to pull him up but it was too much beef for him to handle alone. Dilip came back up and tried to help him but even their combined muscle was not enough to hoist the bull of Punjab.

The bull turned into a wounded lion and started abusing them imaginatively in Punjabi. Jay and Dilip listened fascinated and looked to the GTO for guidance in this troubled time. The GTO shook his head and merely said, "Everyone has to cross." The event was timed and they were running out of just that, time. Jay called up Rakesh and the three of them managed to pull up the inspector, all the while cursing the daily and dairy diet of Punjab police.

They managed to somehow get the snake over the remaining obstacles without any more fuss. This was thanks to inspector Malkeet trying to help for a change rather than just pass orders. It's easier to help than talk when you are out of breath. They finished and were given a five-minute break. Inspector Malkeet miraculously revived and used the break time to cosy up to the GTO.

He was simpering and actually had the GTO laughing. Malkeet would turn every so often during his 'sucking up to the boss' routine and look at the rest of the group. The look conveyed triumph and conviction, conviction that he had nailed the GTO's favour. Rakesh and Dilip looked as if someone had stolen their sister and mother.

They didn't really care much at that time. They had started out feeling tense and excited, and now by noon they were feeling only tense and exhausted, physically and mentally. The GTO looked smug; he had achieved what he had set out to do, put the candidates off balance. There were definitely some who were beginning to get unbalanced. Ramulu was seriously asking Rakesh what the starting pay of an officer was. Jay had never given a thought to that, but understood the logic behind Ramulu's question. It was a question he would have occasion to ask himself often over the next few years, if the pain was worth the remuneration.

They were called back and told to choose a topic and give a short five-minute talk. The topics were of varying complexity. Jay was tempted to pick up some esoteric theme to appear more erudite, but had been cautioned to pick a topic that he was confident about. The idea was to show some mental resilience and confidence while speaking a little sense as well. But that point was lost on the two great North Indian rivals from Haryana and Punjab. They insisted on taking topics that were basically counter to each other's and then addressing a critical audience of two, the rival and the GTO.

Ramulu, who had been brutalized the whole afternoon doing Neanderthal tasks, was elated at the chance to show off his intellectual prowess. Ramulu had never been outside his state and barely spoke English and very little Hindi or Hindustani, that too of the Deccani[6] variety. He announced "To whomever it may concern" that he had been a debate champion in his school. Unfortunately he had been a debate master in Telugu. The talk was to be given in English, and that was not debateable. The topic he chose was child marriage.

The orator from the South started off, "As the great master of Tirupathi say, he who live simply need not fear..."

"Fear what Ramulu? I thought the topic was about child marriage." interjected the GTO.

"Sar I wasu cominga to thatu," said a perspiring Ramulu. His tense mental state was screwing with his diction and enhancing the 'u' endings as in Telug'u'. Inspector Malkeet sniggered softly and things went downhill for Ramulu thereon who wound up his talk with a rapid fire Telugu dialogue which nobody understood. All this finished with a surprisingly good talk by Dilip. Jay had thought his superior English would win the day, but the confidence and clarity of thought displayed by Dilip was more impressive. Jay realized it was not just fluent or bombastic English, but potential that they were looking for.

They were 'broken off for lunch, but after that some of them would have the interview this very afternoon.

The chosen ones of the day were announced. It was the honourable inspector and his dearly beloved Rakesh*ji*; they would have the dubious pleasure of going for the interview that day. Jay secretly hoped they would start fighting and thus get eliminated. The others were relieved; they could now pump the great sons of greater Punjab for information on the interviewing officer's persona. Jay felt a reprieve... he was exhausted!

They trooped back to their barracks and most of them just crashed. Some like Jay who had not been able to complete their shit, shave, shower routine, decided to complete it now, now that the time was ripe, literally. By the time he came back he saw inspector Malkeet carefully oiling and combing out his beard while checking out his profile in the mirror. Sitting there in his undies, with his unrestrained paunch, he looked like some evil genie about to get his wish granted.

Rakesh was sitting on his bed reading from his 'Hanuman Chalisa'[7] very intently. He had also oiled and primped his hair. Unfortunately his tie knot was too fat, it looked like the knot was concealing some appendage. Jay volunteered to help him make the tie look a little less suggestive. A few minutes later, Rakesh finished with his religious reassurances and was now frantically poring through some general knowledge books to brush up on his information. Rakesh had been through this route before and so Jay made a mental note to follow suit.

A couple of hours later the two were back after the interview. The inspector looked triumphant and just could not stop twirling his whiskers while Rakesh looked dejected. The others crowded around him and chorused, "What happened?! *Kya pucha* ?!"

Inspector went to his suitcase, took out a bottle of whisky and said, "*Bahut tagda impression jamaya*. They want a strong warrior type and *oye* I'm one. *Dekh ke pata lagta hai hum dushman ke liye* danger *hain*!" With that the inspector let loose the loud fart he had bottled up. The others left him alone; he was dangerous all right...

Rakesh was feeling aggrieved. The interview had not gone smoothly. He had been given a situation wherein his aged mother and pregnant wife were under threat, and he could save only one at a time. Rakesh had promptly chosen the mother. His logic was simple, "I can get another wife and get her pregnant too, but not another mother!" The interviewing officer had evidently not seemed very happy with the logic. Rakesh kept asking everyone what their answer would have been. He desperately needed to validate his maternal loyalty.

Nobody had the right or safe solution and the rest of the group drifted away. Chib hung back with the inspector and was listening to his master's voice. Jay decided to go to the anteroom and watch the news on TV. He wasn't really feeling hungry, nervous tension does that to appetites.

The anteroom was quite full. There were candidates from other groups, and anyone and everyone was acting as though he was already an officer. The NCO in charge was bossing over some of the candidates, in English at that. He was clearly overstepping his mandate and enjoying his brief moment of power. "No! No sitting here, no talking... Seemed like everyone wanted to be 'officer like' in this place.

Some of the Chandigarh boyz sidled up to the NCO-in-chief, guardian of the carom board and head priest of the hallowed anteroom. They positioned themselves and struck up a seemingly casual conversation with him. The NCO was an old hand and like just about everyone else in Allahabad, could recognize an eager SSB candidate. More importantly, he knew exactly what was on their minds. But he waited, toying with his audience as they circled round him asking the initial superficial questions.

The hyenas got tired and finally asked the NCO, "*Kaise pata lagega* SSB *mein* okay *chal raha hai performance*?" The NCO waited till he had all their attention and then with all the poise of a veteran stage actor he delivered his dialogue in perfect English, "Wait till the fifth day, you will get to know how you performed!" He had obviously been practising and had delivered that line many times before.

Jay thought about the fifth day, lost his urge to watch TV and decided to go back to bed. That night he dreamt he was a salesman in a shoe shop and inspector Malkeet walks in with a couple of girls. Jay woke up in a sweat, not because Malkeet was in his dreams or the girls were, but because Malkeet had been in his dreams...wearing an army captain's uniform.

The next day was another ordeal with the GTO. It was the day when they would have to deal with the half-group tasks. They would be divided into smaller groups and given tasks, giving the quieter ones in the group an opportunity to show off their potential. It did not really go that way as Jay's group had inspector in it with Ramulu forming the third angle of the love triangle.

Malkeet was of the firm opinion that all other races, particularly South Indian, were Slavic races meant for serving superior races like the Sikhs. And thus it was that Jay and Ramulu were reduced to sepoys under the demanding despot that Malkeet had upgraded himself to.

The task they were given, to bridge a gap with logs, ropes and the like, was a complete mess with inspector barking imprecations rather than ideas. He had evidently not heard of General George Patton since he was not from Punjab, or he might have, just might have heard the general's famous line-'Lead, follow or get out of the way'!

'General' Malkeet had promoted himself from inspector to general on taking command of the half group. But he neither led, followed nor got out of the bloody way. Any suggestion by Ramulu or Jay was met by a loud and threatening "*Oye tu meri gal sun*! Listen to me *oye teri maa di*!"

The GTO watched the drama with mild interest. Finally Jay and Ramulu managed to get the task going while the police inspector turned martial ruler kept trying to explain to the GTO how it was not his fault he had useless troops. The GTO had had enough and brusquely asked them to get back to their group post. The rest of the day they had to endure Malkeet's accusing looks and muttered threats in Punjabi laced with inventive abuse. It was also the day Jay was asked to be ready for the interview. All in all it was a bad day.

That afternoon Jay dressed carefully and checked out his reflection from every angle. He was trying his best to see if he looked like a young officer. He thought he looked like some soft kid rather than an officer. He feverishly killed time, preparing for the interview by reading general knowledge books and rerunning the answers he had practised with Col. Bhargav.

Jay went in at least fifteen minutes early for his interview and spent that time sitting erect in what he hoped was an officer like manner. Finally it was his turn and he saw the previous victim walk out. The poor soul seemed to be in a real hurry and had damp sweat patches on his shirt, especially under the armpits. Jay felt an insane urge to walk out and keep walking till he got home. But the danger at home was infinitely more than what he was likely to face inside the interview room. He knocked at the door and asked as politely as he could, "May I come in, sir?"

The effect was slightly marred because his voice cracked. But he was reassured by the kindly voice that asked him to come on in. The interviewing officer was a friendly looking grey beard wearing civilian clothes. He introduced himself as Col. Gurinder Singh from the Guards.

Jay crossed his fingers; of late he had not had very good luck with Punjabis. The interview went off fairly well with only two trick questions. One was about his basketball prowess. He had been asked about the size of the hoop and Jay had managed to put in just the right amount of hesitation to make it look unrehearsed. The interviewer smiled gently; evidently he had not been fooled.

The other question had been about his daily routine which he had not rehearsed. He was about to tell the truth and then remembered his parents griping about it. He told the interviewing officer what he remembered of his father's stories about his own life during his younger days. He tried to water it down, but it was hard to make up stuff when the guy in front of you looked kindly, but had a knowing glint in his eye. Still it wasn't too bad and Jay did not have too many damp patches at least not below the belt when he walked out.

The next day went in a blur. Jay kept analyzing the interview in his mind. This day they had the individual leadership tasks; Jay couldn't decide who would be more pathetic—Ramulu or Malkeet. Someone would be made leader and allowed to choose two helpers to complete the task. Jay had heard that the more often one got called by the others as helper, the more positive an impression he made, showing what Col.Bhargav described as 'sociability'.

Rakesh and Dilip were experienced hands and had made pacts with the others to be called when they were made leaders. The inspector, who was suffering from a hangover, had made it clear that he wanted to be called only twice, and it would have to be Chib or Ramulu who he seemed to have adopted. The inspector had a talent for sniffing out fear and subservience. Nonetheless, fear or not, this was a make or break situation and there was no way Jay could jeopardize the last chance he had to make a good impression.

The task went off okay except he got stumped halfway, and his chosen two Dilip and Rakesh, kept trying to make things difficult by hissing out ideas. And that made him look dumb in comparison to his command. But Jay managed to recover and even earned a smile of what he hoped was appreciation from the GTO. It was actually amusement, as the GTO had seen enough candidates trying to make an ass of themselves by trying too hard.

They got back to their barracks and relaxed. Most tried not to think about what they would do in case they did not get through. Some like Rakesh were very vocal. He wanted to open a Taekwondo school. Inspector*ji* of course would go back to his police work and live off the land and its inhabitants. Dilip was not very happy with the alternative of joining the ACC. That would mean another five years. He would like to get married only after becoming an officer so he could get a pretty 'convent' educated young bride, and time was getting on rather swiftly.

Dilip saw Jay and Ramulu looking a bit confused and clarified, "Officers get better girls..."

Ramulu whispered eagerly to Jay, "More dowries also aren't it *na*..."

Jay was a bit disgusted that this insect wanted to join the Army for money and not for honour, action, freedom and other such unrealistic and boyish desires. He ignored Ramulu and concentrated on what he would say if he had been asked about dowry by the interviewing officer. He had been advised to keep thinking of possible difficult questions and simple answers. But the first answer which came to his mind was 'to hell with dowry, just want the girl'. It was a good answer, just the wrong tone of desperation though.

The next day was their conference and dowry was not on Jay's mind. All he wanted was to get a job and get out of a life that not unlike that of a pet poodle. It was about getting his self-respect back. Though he did not know it, his parent's thoughts were along the same lines.

Jay sat fidgeting with all the other candidates in the waiting hall. You could observe the agony of the death row wait. Jay's turn came after Malkeet finished. Malkeet seemed confident about being selected. He had even performed the victory walk in the style common in the Punjab police, wherein he had fondled and scratched his genitals vigorously as he walked down the aisle waving his free hand cheerfully.

There were two schools of thought inside the conference. The first where there was some doubt about your officer-like qualities, and the other where there was no doubt whatsoever. That being said, the delay in your case being called meant a lot of discussion—either way. Either way it was very excruciating.

Finally it was Jay's turn to walk the plank. He did it with his chin up and a tie that felt like a noose. They just asked him how he liked being here. Jay almost laughed and strangely it was that which gave him confidence.

He had smiled or not (he could not remember) and mumbled something like, "Great! Just wanna join the army and fight for the country!"

They were then seated in a big hall that might have been the cook house at some time. The results were announced in a simple and brutal fashion. Those selected were called out by their 'chest numbers' and asked to stand up while the rejects stayed down.

Jay had almost given up hope when he heard Ramulu's name being announced. He sat nervously waiting for his name to be announced and then it was. Jay was too numb to get up but he did, as a token to his first orders under pressure. The feeling of success was indescribable. He could now inform his father and keep the wolf of humiliation away from his door...just a while longer.

He went out of the hall and saw those who had been rejected sitting with their baggage on the roadside...well actually by the bus stand. The first-time rejects looked dejected for the most part. The repeaters were more cheerful by contrast. They waved out to Jay and shouted out that they would meet in the Army someday. Jay waved back, but some of the other selected candidates did not even spare a backward glance. They were of course the chosen few or so they thought...at least till the almighty medical tests and merit list decided who would get the call letter to join the IMA.

NOTES

1. ACC or Army Cadet College, exclusively for service candidates or soldiers.
2. Minerva Academy is well known SSB coaching center in Punjab that has coached thousands of candidates.
3. Khalsa means pure and a common appellation while referring to Sikhs.
4. The press gangs were recruitment parties used in eighteenth century Britain and other parts of Europe to literally kidnap recruits after getting them drunk or knocking them out.
5. EME or electrical and mechanical engineering corps, responsible for the repair and maintenance of almost all Army arms and equipment.
6. Deccan refers to the Southern part of India. The dialect of Hindustani or Urdu spoken there has its own slang as well as distinctive accent and is a little different from what is spoken in the rest of the country. Often parodied by comedians and in movies.
7. A Hindu religious mantra supposed to ward off evil and ensure success. Draws inspiration from the monkey God Hanuman.

Chapter 5

MEDICALLY FIT

Jay was living in a dream. He had actually got through the interview. He was now on par with his father. He felt really sorry for Rakesh and went looking for him. Rakesh had been through this route and was determined not to give up. He was smoking furiously and cussing as fast and furiously as he could in Haryanvi about the injustice of it all.

Jay and his new-found batch mates, Ramulu and Chib wondered at the logic of not selecting a 'karate kid' sportsman and committed martial race assassin type like Rakesh. They secretly revelled in the ego boost of having been selected over these supermen.

In a newfound boost of officer-like capabilities, Jay and the newly-selected candidates discovered they were qualified to critically analyze the rejected lot's performance. But as the discussion progressed they realized there was an element of luck. Jay definitely felt the hand of Karma, at least three generations of it.

The fight wasn't over and now they had to move their stuff to the barracks where the 'medical' types would stay. Till their medical was over that is, and if it went bad, they too would perch with their luggage at the bus stand.

They hoisted their stuff and moved to the 'medical' barracks. Ramulu followed them still cribbing about the standards that had gone down, "thus necessitating 'officers' to carry loads... unbecoming!!" and continued grumbling in that vein.

They reached the barracks and found a few boys from the previous batch lounging on their charpoys[1] . This lot had superior looks and some even had sneers. They had almost completed their medicals check-ups for the most part.

There was a palpable air of surety that they had at least one foot and several toes in the Academy and thus a toehold or more in the officers' club. They therefore felt an overwhelming urge to exhibit superiority over these civilians, aka Jay and the other newcomers.

It was hard for everyone there to remember that they would all be training together. If selected after the medical, the struggle was still not over. They would have to make it to the merit list. There was still a long way to go, but the martial tendencies of establishing superiority or in this case, seniority, was already in play. The previous batch was acting superior and bossy. They brusquely told Jay and his gang to occupy the beds in a dingy corner.

Jay and the rest acquiesced, they were too fatigued to do otherwise. The last thing Jay did before crashing was to make a mental note to someday rag the sonavabitch who was already wearing a regimental scarf.

His nap was cut short by a visit from their old nemesis, Inspector Malkeet. They were a bit surprised that he had managed to get inside. The inspector was in high spirits; maybe he realized that police was more lucrative. He grinned broadly at everyone and suggested a round of tea. The suggestion was also done suggestively, but no one took the bait, except Chib. Jay ignored the inspector. But Chib positively wagged his tail and gambolled over to the inspector, who took note of this appreciatively.

He said, "Good. *Shabash*! You are now becoming an officer and a man and you need to come with me to Mihirganj for a treat!"

Inspector Malkeet repeated his offer expectantly, but nobody else took the bait, so he picked up Chib (literally) and left. There was some discussion about 'Mihirganj' and whether it was some 'posh' location of Allahabad. Dilip walked in to hear the last part of the conversation; he laughed mysteriously and said, "Chib *saala*! He's too short and too young for shopping in Mihirganj!"

Chib was back an hour later, looking really morose. He went straight to bed and lay face down. The others clustered around.

"Hope you didn't get beaten up?" asked someone.

"Hope you didn't get into a fight?" asked Ramulu who saw the wood for the trees. A fight reported would mean a Chib gone.

"Maybe the inspector raped him..." suggested someone else. The others gave this input some thought, after all it was not beyond the inspector. He seemed to have tendencies of that sort.

Chib refused to answer and the others ignored him after sometime. They would have forgotten the incident had Dilip not revealed it all.

It turned out that Chib had accompanied the great inspector Malkeet to Mihirganj, an old and infamous 'red light' area in Allahabad. This locality had enough prostitutes and sexual congress per square meter to qualify for a world record. Chib had been eager to add his bit...pun intended, to that record. Malkeet being an experienced hand at bargaining especially for love in its physical form, had taken the lead.

"Your tits are too small. There's no way I'm going to pay you hundred bucks for one shot!" he had confidently declared with a worldly air.

"They're too big..," the whore had said, cupping her breasts till they threatened to spill over, "your friend would drown..." pointing her sagging tits at Chib, who was looking at them, fascinated. It was the kind of fascination reserved for the hungry, the depraved and the deprived. Chib was all that.

Inspector Malkeet had walked on to inspect the fleshy wares of the next stall. He was engrossed in checking out the woman, evidently of Mongoloid stock from the North East or Nepal. She was hairless and smooth and the sort very different from what he was used to in Punjab. The Inspector from the land of big, brawny, brave and hirsute people had been fascinated.

The cry, "*Oye Sardarji, apne bete ko to le jao.!*" had brought him back to mother earth.

He had turned and seen Chib still standing entranced by the prostitute who had waggled her breasts at him so tantalizingly. The inspector was a man of action and he did not waste any time. He had promptly caught Chib and slapped him. Once Chib had come out of his trance, he told him to get lost so he, Inspector Malkeet, could get on with his plan and its happy ending. It would give him bragging rights back home in his village.

Chib had also been hoping for similar privileges. But it was not to be and he had to endure the derisive laughs of all the prostitutes in the line, as they unloaded their pent up degradation on poor Chib. He had beaten a retreat and the embarrassment still had him red-faced and almost in tears.

There were very varied reactions from the audience of possible future officers. Some guffawed loudly, some were sympathetic, and some like Ramulu completely missed the point.

"But he wasu justu looking at the whoreu..." he said, his agitation again thickening his Telugu accent which mandated the addition of 'u' to the end of every word... but no one was interested by then.

They were all worried about what esoteric medical disability the military hospital was going to use to disqualify them. A new cause for tension had replaced the euphoria of the last few hours. Jay decided to break out the Old Monk rum his father had thoughtfully given to cater for emergencies. This sure felt like an emergency and the rum soothed their distressed souls effectively. The next morning Jay woke up with a mild hangover. Chib seemed to be in similar straits. Dilip had a few words of wisdom.

"Drink a lot of water." And he continued when he realized he had their undivided attention, "it helps wash out the pus cells"

"But I thought pus cells are there in even the normal male man's urination!" suggested Ramulu hesitantly.

"If you want to be that normal male 'man' who gets rejected its fine by me." said Dilip nonchalantly. Jay was the first to reach the water cooler, closely followed by the others.

An hour and several litres of water later Jay felt he would drown if he drank any more. He was relieved when it was finally his turn to give a urine sample. The rest of the medical check went off fine. Dilip was temporarily rejected for some reason and given a date to report again for a check-up. Rumours were rife and the most popular story was about the candidate who was rejected for giving cocky replies. The candidate was being checked for piles and haemorrhoids. The conversation as legend has it went as follows:

Doctor: Take off your pants and bend!

Candidate: Checking for officer—like parts?

Doctor: Shut up and bend! (Strike one)

Candidate: Cleaned myself up just for this. Everything okay?

Doctor: Stop talking...Do you smoke?! (Strike two)

Candidate: Why? Is there smoke coming out!?

Doctor: Get out! You're rejected. (Strike three)

The story was pretty old and stretched the bounds of incredulity, but still retained its effect. The principal effect was to ensure absolute discipline and total obedience to all commands, medical or otherwise. And this meant subservience to anyone remotely military in appearance. The lowly denizens of military hospitals from sweepers to nursing assistants enjoyed their moment of authority and importance. You would never find a better behaved group of humans anywhere.

Jay had an anxious moment when the lab assistant called him back and asked him to give another urine sample. He dared to ask, "Is everything all right...?" The nervousness in his voice puffed up the lab assistant.

"Did you mix water in your urine sample?" the lab assistant asked accusingly, "We have seen it all here so don't try any more tricks!" as he handed him another test tube. It seemed that the rivers of water Jay had drunk had flushed his bladder so thoroughly that even the normal percentage of pus cells was not there.

Two days later their tests were over and Jay was told that he had cleared the medical test too. He had been worried about his weight and the urine test. Luckily he was within the ten percent range of his weight limit and so managed to scrape his paunch through. And now there was nothing between him and the Promised Land.

The next few weeks sped by while young lieutenant designate Jay basked in the quiet glow of victory after a hard fought battle. His parents basked in the reflected glory. That part was not really clear; was the glory reflected off Jay...for having studied hard and gotten through the exam on his own... or was it reflected off his parents for having provided the magnificent gene pool that ensured his acceptance to the IMA. Either way nothing was too good for Jay.

His father had promised him a motorcycle on joining the IMA. That was sadly postponed to 'after getting commissioned'. But Jay got his first pair of readymade jeans, among other things. His father also bought him a new blazer and shoes, in a rare fit of extravagance. After all, nothing was too good for a son who had put the spirit of his ancestors to peace, and helped Shankar senior hold his head high in the club. He was looking forward eagerly to bumping into Col. Sharma or some other proud parent of that sort. Victory has a sweeter taste while comparing successful offspring.

NOTE

1. Charpoys is an age old form of a cot still used widely in the rural parts of India. It was also part of the bedding in the barracks consisting of a wooden or bamboo frame with netting made of coir plaited rope. The bottom part of this contraption has a horizontal series of rope twining to aid tightening. The charpoy has served numerous roles over the last few centuries and still does, including that as a litter for dead bodies and a stretcher for the wounded.

BOOK II

BOY BECOMES MAN - I TERM

Chapter 6

MUKHYA DWAR OR IMPORTANT GATE

Jay was very, very excited about the whole adventure. He had spent considerable time poring over faded black and white pictures of his father's days at the IMA. There were pictures of him and his old course mates, some of whom Jay had met. Some, his father had informed him, had been killed in the 1965 or 1971 war but there were others who were doing well.

The doing well types, not surprisingly, had sons who were in the Army also doing well. Jay had been suitably impressed by the pictures of tough looking young men in various action poses and smart clothes. He had not been impressed by his father's heavy hints about successful sons. But all that did not matter now. He was here in the IMA now and he was a successful son.

The clothes they wore at IMA were interesting—dungarees, web equipment, striped sports blazers, mess dresses and so much more. But but the most interesting aspect was the fact that all the guys in the pictures were smiling. They must have had a 'bloody good time' as Col. Shankar often reminisced about the 'good old IMA days'. There were moments of doubt when he saw the old man's IMA dressing gown with holes in the elbows and knees. This was because they were 'gentlemen' cadets and not just cadets. They had to be correctly dressed even during punishment parades at odd hours. It had all sounded so British and 'propah' and Jay could not wait to join the club.

Jay's sister was also eager that he join so she could regain the favourite child status. And so it was that he reached Dehradun a day in advance and had to stay in a local Infantry brigade[1] at one of the battalion's messes. The CO of the

unit was a family friend and assured the Shankar's that their son would be safely deposited at IMA at the appointed hour. That night Jay had his last drink for some time to come in an officer-like manner.

The young officers who were dining in the mess treated him as their own and were happy to see a creature junior to them in the mess. They got him drunk while telling him stories of IMA. The stories were flavoured by alcohol and made the Academy seem like one happy public school full of naughty boys, stupid seniors and crazy instructors. Of course none of whom could ever get the better of the smart gentleman cadets that the storytellers had been. And of course Dehradun babes were crazy about the gentleman cadets Jay could hope to score on his very first out pass, or liberty as it was also called. He would get to why it was called liberty soon, but that night Jay went to bed a happy carefree kid with unreal dreams.

The next day as Jay was driven through the gates, he found himself struck by two observations, both of which sent a slight chill down his spine. The first was that nobody seemed to be smiling and the second that people seemed to be functioning without any care for the weather. There were a few cadets from the ACC[2] wing racing about in shorts and vests. The chills took his mind off his hangover and to the fact that it was biting cold outside. He was after all joining the winter course. It is still a subject of heated debate—which was the better term to join or pass out of—summer or winter. Jay would learn that the Army was all-weather.

The main gate of the IMA was called the 'Mukhya Dwar' which simply meant 'Main gate' or could also be interpreted as 'important entrance'. For Jay, it was one of the most important entrances of his life. But he was also worried about his mode of entrance. The antipathy towards *fauji* progeny had been made aware to him during the SSB. The last thing he needed was a reawakening of those emotions on his very first day by entering in a plush army vehicle. The two thoughts boiled down to the single realization that he would have to prove himself. And that added to his worries.

He was worried if his dress was too over the top. He worried his haircut was too long. He had mentioned the haircut to the young lieutenant who had volunteered to escort him to the IMA. The others youngsters at the mess had all been volunteers. Everyone wanted to reminisce and watch the fun as the new sheep were rounded up and culled. There had been much mirth when Jay asked if his haircut was short enough.

The answer was enigmatic enough, "Don't worry son, your hair can never be too short in the bloody Army!"

He realised the jeep had come to a stop. Lt. Raja Chetri, who had chosen to drop him off, got out smiling broadly. "Alright boy, the ride is over and the real fun will start in about six hours from now when it gets dark."

Jay got down, a bit self conscious about his brand new blazer and school tie, chosen because it looked sort of regimental. They were near the drill square in front of the Chetwood building, the landmark of IMA since its founding. He saw an orderly line of cadets wearing NDA blazers and the cadet appointments. He asked Lt. Chetri where the rest of IMA was. He was informed that the Cadet appointments and the ex-NDA types along with the direct entry types like Jay were reporting a week earlier than the rest of the cadets. With that Jay's stuff was dumped with the rest of the luggage near a group of bewildered looking civilians. None of them wore ties, coats or had had a haircut.

The group perked up when they saw Jay. Here was someone who seemed to be of a higher pedigree than them. After all he had come in a military vehicle, had his hand shook by a regular army officer and his trunk and 'hold all' carried down by army soldiers. The crowd looked expectantly at Jay. They waited for him to say something and reveal the mystery of what was to come. Jay was just as clueless as and possibly more ignorant than them. He smiled at them awkwardly and asked if this was where they were supposed to report. The answer came in the form of a cross between a shout and a scream.

It was from a tiny drill instructor from the Gurkha regiment. The little bantam asked them to leave their luggage, get their documents and get into three lines. This was conveyed to them in a tone out of proportion with the size of the delivery vessel. It was truly amazing how a little goat-like Gurkha could bellow like a bull. His name was Subedhar Tamang something and he was something of a terror, as they would all soon learn.

The crowd of civilian candidates was unnerved by the authoritative tone of Subedhar Tamang. He was wearing the rank of a JCO or junior commissioned officer, consisting of two stars and a stripe. This was often mistaken to be an officer's rank. The new cadets jumped to it and formed three lines as requested with minimum fuss. In that instant they had taken their first true step towards soldiery and became gentlemen cadets.

The next few lines were made up of ex-NDAs. They were lined up obediently too but every so often they looked over at the line of DEs or direct entries[3]. The look was not unlike that of a caged predator upon sighting a prey within reach but separated by the cage. In this case, the cage was the presence of instructors and cadet appointments.

The DEs did not know it at the time but one of the reasons they were reporting a week earlier was so that they could get a chance to settle down before the rest of the Huns descended on IMA, thirsting for virgin blood. This was metaphorically speaking but the social situation did not look very different.

The cadet appointments[4] also reported early, supposedly to learn the cane drill and help the newcomers adjust. The real reason was to keep the newcomer DEs and ex-NDAs in line...While helping them orientate themselves with various shortcuts and loopholes. The Academy's intentions were noble but not that of the inhabitants, especially the appointments.

After the paperwork they were told which companies they had been allotted to. Jay found himself assigned to Singarh Company as part of Bhagat battalion. He knew a bit of military history and had an idea what all this meant. The companies were named after famous battles and the battalions after famous generals. He could explain this to some of the others who had never heard these names. What he could not explain to his company mate, a Bihari named Sinha, was why there seemed to be nothing named after Nehru or any of the Gandhi's.

They reached their company, and Jay was not impressed. They were just buildings with tin roofs, like any other hastily constructed barracks. Jay was disappointed because what he had imagined, based on what he had seen in his father's pictures was a quarter of a century old. Since then the Academy had grown in strength but not in infrastructure.

They were all lined up once more and the cadet appointments began to show their true colours. Their stiff attitude in front of the DSs or directing staff as the officer instructors were called was very different and the new DEs felt it.

The company line was divided into three platoons. The rooms of each platoon were laid out in squares with the bathrooms and latrines shared by two platoons, being laid out in between. Also in between the squares, known as the quadrangle, was an open area with a cycle stand, the odd ornamental plant or tree and enough place for the platoon to fall in. It was the scene of much grief with the ground drenched in much sweat and some blood.

The newcomers were assembled in the quadrangle and everyone was feeling a bit odd. The oddness was because the atmosphere was now that of a slave market. They no longer felt like gentlemen cadets and future officers. The ex-NDAs also seemed to share the mood. They were sized up, scrutinized and did not measure up, going by the expressions of the appointments. The results of their evaluation were disseminated by a very fit and tough looking young man who was the BUO or battalion under officer.

Actually they were all tough and fit looking. Jay had never seen such an absence of paunches in his life. He was brought out of his train of thought by the stream of invective the BUO was throwing at them. It was invective but not really unkind sounding; it was actually part of the official welcome speech.

It went something like "Welcome you fuckers and God knows why you sorry assholes decided to join the Academy...and shame the bloody army" and more in that vein. The welcome was also mixed with a lot of dos and don'ts. They would hear a lot more definitely dos and positively don'ts in the future. The tone of the dos and don'ts became a bit more strident. The DEs were getting their first introduction to an ML or a 'moral lecture'. Jay would soon learn the importance of moral lectures and accept the fact that a major part of his life would be consumed by them.

The BUO, having finished his spiel, went off to address the newcomers in some other company. Jay looked around and saw the ex-NDAs looking bored and unimpressed, having had three years of military training and rations. The DEs of course were thoroughly cowed and somewhat overawed as the ex NDAs reputation preceded them. The ex NDAs on the other hand were sort of underwhelmed. The fact that they were now once again at the bottom of the food—or actually ration chain—had not really sunk in. They had what could be put mildly as an 'attitude problem' and the cadet appointments knew that.

The cadet appointments intended to ensure total obedience and dominance and they had to do it fast. The DEs were collateral damage. The CSM or cadet sergeant major set the tone for their future and announced 'order fall in' would be at six and muster again after dinner, at nine.

The newcomers were distributed into their platoons and handed over to a JUO or junior under officer. Their new master was just about five foot four even though he had puffed-up like some bantam rooster. He strutted around them looking real important. They were four DEs and about ten ex-NDAs in the platoon. The latter group quietly picked up their stuff and went to the unoccupied rooms. The DEs stood about looking lost. The JUO, whose name tab announced him as Akshay Kumar, introduced himself.

"I am Junior under Officer Akshay Kumar, Hunter squadron and Sainik school (of some obscure place)..." He paused at this and looked expectantly at his flock. They looked back at him blankly, as most of them had never heard those names before and had no idea of their significance. Akshay boss was disappointed, but he promised them, (he paused here with a theatrical laugh) they would get to know him. They later learnt that the little JUO was from one of the toughest military schools and then one of the most brutal squadrons of NDA. If he had belonged to any other species he would have been called pit bull. Life would not be easy under him. He was suffering from the mother-in-law syndrome—'Do unto others as it was done to me when I was young'.

The immediate future looked very bleak.

NOTES

1. An army fighting brigade like this, the infantry brigade, usually consists of three major battalions apart from other supporting arms/service minor units
2. ACC or army cadet corps was a wing of IMA where cadets from the ranks or ex soldiers/sepoys after selection underwent training and completed their graduation.
3. Direct entry was the name of the type of commission and denoted the fact that they were entering after graduation
4. Cadet appointments were appointed at every level from platoon and company to battalion and academy level to help administer and teach leadership and or management skills. It was also an indication of the cadets potential and his merit and was fiercely competitive. They also carried canes as symbols of their authority.

Chapter 7

THE GENTLEMAN DEFAULTER

The punishments at Sainik Schools and NDA were copied by lesser sadists in other Indian military academies. It was not to be confused with or referred to as ragging—that was some disorganised and effeminate crap followed in civilian institutions. This school of thought has not changed since military academies were invented.

They (the military) swore punishments made boys men and ragging just made you disgusted. Those who had survived and graduated from these institutions were ever ready to try out this theory on others. It did not matter how disgusting the resultant feelings, the cycle never ended.

JUO Akshay Kumar was hot and ready to put the theory to the test...He deftly singled out the man on Jay's right with his cane[1] . Oh yes JUO Akshay Kumar had been give a symbol of power and authority by the Indian Army, and 'By Jove' he would use it. And so it was that he prodded Cadet Anant GRP Kumar with his two and a half foot piece of wood and steel and asked him if he had a good handwriting.

GC Anant Kumar thought it over and asked instead, "Where is my office?!"

There was hushed and pregnant silence while the first termers looked anxiously at JUO Akshay. JUO Akshay too looked a mite anxious and confused.

Cadet Anant Kumar was undergoing what is usually and delicately termed as 'inner turmoil'. Jay noticed for the first time that Anant Kumar's trousers had dirty knees and he sported a smeared tika on his forehead. In fact he looked like he had already been through a 'punishment'. Jay and the others looked at him with awe, while JUO Akshay Kumar looked at him with irritation.

JUO Akshay Kumar would have to sort this chap out. Questions...doubts were not encouraged in the academy. Disobedience was a fire you did *not* want spreading.

Anant Kumar had arrived at the Academy with a light heart and the confident assumption that he had managed to get a government job. True... that job may not be as lucrative as other central services but he would still be a class one gazetted officer and he would still have status and perks. A little dowry might also be there to sweeten the pot when he became an officer.

He had been sadly shocked when he learnt he did have a status, and that was something in between a slave and bonded labourer. He had demanded of the first cadet he had come across, "Where's my office?"

This was within a few minutes of being told which company he had been placed in. Anant Kumar had unfortunately thought he was joining IMA as a company commander and not a lowly insect called cadet. It was thus 'General' Anant Kumar had made the grave error of accosting a cadet, who was wearing a strange leather band on his wrist.

This leather-band wearing cadet was a CSM or company sergeant major, one of the lynchpins of discipline in the Academy...indeed in almost all armies. In Indian military academies, the CSM was feared and revered in varying measures depending on individual tastes. GC CSM, R.S. Gaekwad was a true

gentleman and thus did not take umbrage at the uncouth form of address by a lowly first termer. Nevertheless to ensure decorum, he gently suggested GC or gentleman cadet Anant Kumar to get on his knees and extend his hands up in supplication.

"Someone will be by shortly to answer your prayers and allot you an office." But Gaekwad saw that GC Anant was not ready to accept suggestions as yet. A gentle suggestion is a rare thing in the academy. Since this GC was new, Gaekwad continued, "That's how things get done around here, you believe in God don't you?" He realised Anant was still looking a bit sceptical so he added in a conspiratorial tone, "Believe me, I'm a CSM and I know how things get done here, you have to please the seniors...I mean elders here by following tradition!" This last was added, pointing at the drill instructors who were roaming around looking very authoritative and impressive.

Anant Kumar believed in God and had never heard of CSMs. He was not sure about this kindly-looking GC who seemed to be knowledgeable, but decided to do as suggested. It was the last time anyone would suggest anything to him...after this they would just tell him.

After ten minutes, Anant was still on his knees with his hands stretched out to the heavens in mute supplication for an office. The office allotment letter did not come. However many III[2] term GCs, who had heard of this young office seeker, had come by to amuse themselves.

Anant Kumar was a proud man. He was proud of his caste, being a Brahmin, education, he was an engineer and his heritage, his grandfather had been a freedom fighter. He sensed he had been tricked and thus got up in a huff and went looking for the cheating CSM. He found him in the midst of other cadet appointments in the anteroom playing billiards.

"Where's my office?" Anant cut to the chase.

The others laughed uproariously at the question. They had already heard the story. CSM Gaekwad had calmly turned around and told him to report to JUO Akshay Kumar of the first platoon, where he would get further instructions regarding his office.

And so it came to pass that Anant Kumar was standing in the middle of the dirty quadrangle with other I termers and some II termers, all looking at him with a mixture of awe and anticipatory dismay. JUO Akshay ignored the office question. It was too preposterous to even try giving a sensible response. He was afraid he might lose his composure and do something violent. He had been relegated by a term for manhandling a junior in NDA. He did not wish to pay again for a momentary lapse of reason with six months of his life. Especially now, when the end, and his commission was in sight.

He controlled his irritation and said briefly, "Shut up and stand with the rest of your course mates!"

This was delivered with enough menace to shut Anant up and make him join the other I termers standing meekly. Anant made a silent prayer to his dead freedom fighter grandpa to give him guidance and God to save him. He did not know that even the freedom fighters had avoided messing with the IMA during India's freedom struggle.

JUO Akshay Kumar was not done. He still had to maintain a certain level of terror and psychological high ground, and that had to be done from day One.

The worthy JUO stalked his inexperienced victims with a predatory look. It did not require much effort since they were somewhat huddled together. Unfortunately his height did not give him the kind of predator's look he had hoped for. Okay, so his walk had feline grace but it was of the domestic cat type rather than the tiger on the prowl that JUO Akshay had hoped for. The ex-NDAs were not impressed. They had seen worse. But it seemed to satisfy Akshay the tiger's predatory instincts, and he pounced on the GC next to Jay. This was a gentleman named Arjun Shinde.

Arjun Shinde was a good-looking prototype for the classic Marathi warrior. He was tall with a classic profile, well-built, an erstwhile wrestler, and a sincere Ex NCC cadet with strong martial instincts. These instincts had made him prepare for a career in the Army from a very early age. What it had not prepared him for was JUO Akshay Kumar. GC Arjun Shinde was everything Akshay Kumar was not and so he was irked. 'Hell hath no fury like a senior irked', is something every military creature knows, and Arjun Shinde would know it too.

"How good is your hand?" asked the JUO mildly.

Arjun was lulled by the deceptive tone and replied, "Very strong sir, I was a bodybuilder..."

"I was referring to your handwriting, you idiot!" said Akshay Kumar, "One of you will have the honour of being my understudy..." He looked at the other two, but could not see any eagerness to claim the honour he was offering.

Anant Kumar added to the aggravation "What duties would we have to perform?" He still held a hope that the duties of an understudy might get him an office.

Akshay Kumar realized if he carried out any more banter with these lowly civilians he would definitely lose his edge of terror. And terror was the key to keeping an undisciplined rabble under control. So without wasting any more time or breath, he invited the four I termers to get down on their hands and give him thirty pushups.

Happily for Akshay Kumar, GC Anant Kumar was unable to do more than ten. Anant was about to get a taste of what the freedom fighters had gotten during the days of the Raj, some from the alumni of this very Academy.

But before Anant's grandpa could start rolling in his grave, a runner[3] came summoning all JUOs to the company commander's office. This forced Akshay to put off his plans for an introductory 'fuck session'.

And he promised to complete this at the very next opportunity he would get in the future. "You fuckers better improve your reactions or I will give you such a fuck session you will not realise which way is up!" He was thrilled by the apprehension and confusion he saw in the eyes of the I termers when they heard the term 'fuck session'. The apprehension and confusion was natural when someone hears a term which implied aggressive sexual congress in a place meant for training defenders of the nation. But JUO Akshay thought he had scared them and was happy.

"See you fuckers when I get back, now go and get settled up in your cabins!" He looked at Shinde and continued, "You take that cabin next to mine...you are now my understudy." He then tore himself away from his slaves to answer the call of his master.

Jay was sweating, dirty and without a clue what his next course of action should be. It had been only four hours since he had left a world of too few jam sessions for one seemingly overstocked with 'fuck sessions'. The others were in the same leaky boat...

Shinde said, "Let get an empty cabin and get ready for the next fall-in."

Anant murmured something about wanting an office not a room. But no one was interested and they drifted off looking for vacant rooms. Jay and Shinde helped lug each other's trunks to their rooms. Anant seemed to have vanished. Jay had just got in the last of his stuff when an ex-NDA II termer came and told him succinctly and impersonally to 'fuck off'. To add to the injury, he and Shinde had to help the second termer move his stuff into the room. Jay finally found another room where he promptly fell onto the bare cot and sank into a dreamless sleep. It was evening when he woke up... someone was kicking his cot.

Jay woke up saw the quadrangle lights shining through the doorway and realised he was late. He shot up, wide awake now and remembered he was supposed to be at some fall-in. He remembered the fall-in but not where and why. He realised his memory problem could make the difference in his stay and the comfort level thereof.

He was not very wrong. He was sent back after a few moments and some pushups to wear a coat and tie. That wasn't the end of it...He got back to his room to find the entire block in utter darkness that threatened to eclipse his sputtering euphoria. It took some time before he got a candle lit (God bless mom for putting in all that useless stuff).

He managed to find his coat and a tie, but the fates were cruel that day and a few minutes later, a II termer opened his door and looked at him unemotionally and asked him if he had any OP shoes.

Jay asked him innocently, "You mean like these..." showing him the shoes his father had bought him just before leaving.

The reply, "Yes, thank you. Now you can take them off.," was the reward for his candour.

A few minutes later, Jay was rummaging again for another pair of black shoes. He was out of luck because all he had was two pairs of extremely snazzy and wholly unsuitable shoes, being of the sporty or casual appearance. The II termer had coolly walked off with his shoes. He had been very polite but there had been the undertone of menace. There seemed to be a lot of that... 'Maybe that's why they are called gentlemen cadets', thought Jay wryly.

Jay wore his tie and blazer and hoped that no one would notice his brown moccasins. He rushed back to the dining hall praying he would not have to enter while everyone was seated and get noticed again. He reached and was most definitely noticed. All the newcomers of the battalion were standing in orderly rows outside the mess hall with their respective platoons and companies. The steps in front of the battalion mess were occupied by the BUO, the BCA or battalion cadet adjutant along with the whole host of CSMs and other lesser devils with authority.

Jay was spotted immediately and snared. "Come here you bugger! Where do you think you're going in such a hurry?" said one of the JUOs who had been lurking in the fringes looking for some diversion. He was feeling left out of the high powered conference of other higher appointments at company and battalion level.

"Sir, I had gone to get a coat and tie..."Jay tried to save himself, but his assailant was already looking thoughtful. Later Jay would learn that any time a senior looked thoughtful he was contemplating nasty options for some subordinate. The subordinate lined up in the senior's sights right now was Jay.

The JUO was a particular sadist from another company and not really concerned with game from other companies. But the other appointments were busy so he took his time pondering. He ended Jay's suspense by asking him to get rolling. Jay knew what 'rolling' was from his dad's stories, it was short for front rolling. One had to roll like a ball literally and for those who were not proficient, this could be quite painful and potentially damaging to the neck and spine. Just as Jay was assuming the position and kneeling down with his palms on the road, he heard one of the other appointments say, "Not that, he's new...might get hurt."

JUO Akhilesh obliged by making Jay get on his hands in push-up position. He asked him to remain like that while the eager JUO went around looking for other victims. Jay just thanked his stars no one had noticed his moccasins. 'This push up position must be more popular that the missionary position in the army' Jay thought to himself. He had lost count of how many times he had gotten into 'push up position' since that morning.

It was about an hour before they were allowed to move into the dining hall. Jay was beginning to think that he had been forgotten, when he was told by one of the other appointments to get up and join his platoon. He joined his company and was pleasantly surprised to recognize a familiar face. It was Ghai; they shook hands hurriedly as they went in. They entered the huge hall with its high roof and rows of tables. They found their seats, picked up their plates and joined the line. Ghai kept cribbing as they served themselves and went back to their place. "Feels like a bloody *langar*[4] ...they should be serving us at the tables, this is so bloody unofficerlike!"

The table only had the rice, *chapattis* and salt and pepper shakers. Jay and Ghai found themselves flanked by two appointments. The one with the black shoulder boards looked at them dispassionately and told them to get lost. They did so with alacrity that would have gladdened the hearts of their long suffering sisters back home.

Jay and his companions were but a few hours old in the Academy but their survival instincts had kicked in. This was after all the jungle and it would be some time before they became Tyrannosaurus Rex. They later learnt that they had narrowly escaped. The blackboard shoulder was the ACA and on top of the cadet food chain. He was also not in a good mood since he had been ticked off by the IMA adjutant who was way above him in the pyramid.

The ACA gave vent to his ire by calling a II termer and asking him to make an announcement. "All first termers to report to their JUOs immediately after dinner!"

The I termers heard but did not understand, and even if they did they weren't in the obedience zone as yet. They were not yet house broken, so to speak, and had had a long day. The ACA and his eager assistants knew that and did not care. They only cared to remember that they had been through similar processes and survived. Hence it was now their duty to hand down the whip to the next generation. It was also morally satisfying, if incorrect.

"Spoons down, you pussies, fall in outside right now...the last defaulter will suffer me personally!!" yelled an appointment after ten minutes.

There were incredulous looks, and then there was the stampede for the door. No one wanted to be the ACA's pet project. But it was the rule of nature that someone had to be last...and unfortunately it was Anant Kumar. The last they saw of him that night was Anant eating his food calmly while surrounded by a bunch of screaming cadet appointments.

NOTES

1. Persons below Lieutenant Colonels rank and JCO's are authorized to carry 30 inch long whippy canes with a bulbous end with a regimental crest and steel tip on the other. Mostly carried by drill instructors and cadet appointments.
2. First, second and third term is being denoted as I, II and III, simply because that's the term badge they wear on their uniform lapels and easier for me to relate.
3. Runner, messengers were called runners, in earlier time's men or boys carried messages on foot. Still considered more handy and reliable than phones and other new fangled communication methods.
4. Langar is colloquial Hindustani for cook house or common kitchen.

Chapter 8

NIGHT PT

Jay was face down in the grass. He was breathing hard. They were all breathing hard and sweating, and yet they were freezing. It was their third day. They were dressed in varying styles and shades of white PT clothes. Most of the white was now almost khaki. And most of this had happened in the last thirty minutes.

After being evicted from the dining hall, yet again, they had 'fallen in' rows that were remarkably orderly in spite of them being just a few days old in the Army. The rest of the cadet appointments had come out of the dining hall with a defiant looking Anant Kumar in tow. It did not look good. Anant had gotten them in trouble yet again.

Their worst fears came true. Half an hour of haranguing left them numb with cold and despair. The main talking point was about obedience being the prime virtue in the Army. It was closely followed by the definite and immutable fact that the sins of one would mean punishment of many. The entire herd of I termers, raw as they were, could sense exactly where this conversation was headed...nowhere good. Nowhere good...in this case the PT ground and they had been given five minutes to fall in, in PT dress.

They 'fell in' again. Jay had lost count of the number of times he had 'fallen in' in his brief career with the Army thus far. But even this short period ingrained a deep sense of aversion to 'fall ins'. But 'fall out' could be just as bad; they would all learn in due course.

A headcount was taken and they were found one short. There was much swearing, mostly in Hindi. Jay learnt that night that use of abusive language and general swearing in Hindi, Punjabi or any vernacular in general, indicated a heightened level of purpose. Indeed this was true of the officer corps in any Army and was more pronounced in the Academy.

This was partly because of the idea that officers should only communicate in English. Vernacular indicated an attempt at familiarity and emotional turmoil, both especially discouraged during official communications.

Right now there was a lot of official communication going on at the PT ground, albeit garnished with foul terms. This was enhanced as various appointments struggled to get their minds around a single fact. There was one first termer missing from a 'fall in'. A fall in that was desired by none less a personage than the ACA. The issue was of direct defiance of just one individual but could infect others. This insurrection had to be dealt with an iron hand or..."the whole term would be fucked up," as third termers put it. The I termers would have begged...yes literally begged to differ.

They seemed to stand there forever while the JUOs checked the platoons and gave the report to the CSMs who then compiled the company's state and gave a report to the CUO or company under officer. The CUO gave the report to the BCA who compiled the battalion state and gave a report to the ACA.

The whole process was derailed from the expected martial efficiency because of the bitter cold. Most of this mathematics was done by understudies who usually accepted the monotonous drudgery of actually making out the neat charts and parade states, accurately. But they were I termers and the current lot did not 'know their arse from their elbow'. This was a hundred year old statement and held true especially today.

All this made the atmosphere very volatile. The aggravation was piling up, tempers were rising...it was but a matter of time, before the III term appointments lost their cool enough to ignore strict orders to go easy on the I termers. The flash for the explosion was given by the discovery of the missing cadet's identity. It happened to be none other than the now infamous GC Anant Kumar. He was forcefully escorted to the fallin.

A duty officer making his rounds heard the commotion and came to investigate. There was Anant Kumar looking serene in his pyjamas while at least four cadet appointments yelled at him simultaneously. Jay's spirits were lifted by the sight of the duty officer. He knew I termers were usually treated with kid gloves for the first one month at least. Ragging or hazing of any sort was strictly prohibited. But training or punishment by seniors, especially appointments were not discouraged...in fact in some cases this was actively encouraged by the officers. It toughened up the cadets, made them attuned to obeying commands even under stress.

The last reason which was often quoted by the cadet inquisitors was to train their victims to withstand the rigours of interrogation, if captured. All this had been conveyed quite forcefully if not convincingly. For added measure it was garnished in colourful and unforgettable language, to the new cadets. As they stood shivering in their thin PT dress, Jay hoped he was just trapped in a bad dream and all this would disappear when he woke up. But his dreams were shattered when the aforementioned duty officer did not rescue them in the way he—and the others—had hoped.

"What's happening? Trying to entertain the I termers...forgotten your briefing?!" the duty officer mildly asked.

This was deceptive since this particular gentleman, Capt Jeevan Singh, had the reputation of being a cold blooded and unpredictable bastard. He also had stamina, taste and inventive spirit when it came to 'sorting out' cadets. He held the current record of punishing cadets for a record nine hours without repeating any of the punishments. The BCA was his fan and thoroughly terrified of Jeevan Singh.

"Uhh sir, we were just having muster fall in. But sir...one cadet was acting funny...did not want to join his course mates in the fall in!" blurted out the BCA.

"Well, I guess then you will have to teach them the joys of course spirit, won't you?" said Jeevan the terrible. "I want to see them sweating when I return after one hour or you'll all be bleeding from every hole..."

The collective spirit of the I termers which had revived when the duty officer made his debut, once again sagged. They looked at Anant Kumar, the cause of all their suffering. They hoped the minions of military discipline would unleash their frustration and bruised egos on Anant Kumar and spare them. But the BCA and his cronies looking triumphant and unfettered, turned towards the huddle of I termers wishes.

This was yet another management lesson to be learnt in HR or human resources. When faced with opposition, make an example out of the path of least resistance. It was an old concept formalized by the Chinese in times old and new, "Kill one and frighten a hundred others!"

The idea had been modified by the British and their public schools for potentially rebellious groups. And it had worked remarkably well for the last 100 years or so and was just as effective today. Jay would have testified to that if he had been asked, and if his heart had not been beating so fast.

It was not just his heart but those of the whole group that was thumping away. The adrenalin was not just because they were like cornered rats. It was because of a variety of emotions that prevented them from fleeing, as per instinct. Some had no other means of fulfilling their family's aspirations. Others had dreamt and planned of nothing else but the army as long as they could remember. To drop out or to be kicked out now would mean that they could be shamed forever... depending on the community and area they belonged to.

Anant Kumar did not have financial and cultural constraints on him to prove his 'machismo' either way. But his indifference was resulting in a lot of his comrades enjoying the dubious pleasures of the military machismo. That did not bother him. All he had ever wanted was his office and the job they had promised him. Surely they would not torture everyone here for that. He was here; his grandfather who was luckily in heaven, had been a freedom fighter. Surely they would not execute colonial techniques on grandsons of freedom fighters.

That was what Anant thought anyway. If he was a child of the freedom movement, his oppressors were the products of a well-oiled public school system that was had defended the system. If Anant had read 'Tom Brown's School Days' he would have known what he was up against. But he had only read Gandhi*ji*'s works which were not particularly helpful as a survival guide in a military academy.

Anant was an intellectual who would have been an asset to any free thinking and or profit-making organization. But unfortunately he had, in a fit of romanticism instigated by his politically-minded father, followed the road less travelled by his community mates.

He had joined an organization that expected you to prove yourself physically and then establish yourself academically. It was his civil servant expectations that had brought him thus far, but being a servant was not part of the plan and definitely not being unpaid labour. He was also unaware that 'freedom fighter' was a derogatory term in the Indian Army. It was reserved for the rebellious sorts or those who chose to question orders and were thus unable to fit in.

GC Anant Kumar therefore decided to adopt the posture of civil disobedience...like grandpa. He did not realise the shit storm he was kicking off.

The cadet appointments were children of independent India and the British system of discipline. They firmly believed that condoning disrespect is fatal in the long run for all parties concerned. Institutional memories also reinforced the dictum that indiscipline was infectious. GC Anant Kumar was one of the first symptoms of the infection called 'showing doubts'.

Showing doubts referred to the acceptance of orders with reservations. This was considered the bedrock of military discipline and good order. GC Anant Kumar was threatening that very foundation. The cadet appointments were eager to show the I termers the folly of challenging authority. But the eagerness was tempered with pragmatism. If Anant refused to obey orders in front of the whole battalion, there would be consequences. The other cadets might follow suit; in short, mutiny. The company officers might decide that the cadet appointments did not have what it takes to maintain discipline and 'de tab'[1] them.

The BCA had considered all this and hence decided to make Anant the focus of the I termers' hatred. The Chinese system of 'kill one frighten others' would not work at this point of time. It was going to be the Indian method of 'shame one and anger the others'. Accordingly, Anant was made to stand at the head of the group of shivering I termers. The III termers then went to work with singular efficiency.

The 'youngsters' were divided into company wise groups which worked out to roughly three groups of 12 cadets each. Orders were given and the PT began in earnest, starting with 'jumping jacks'. They were soon sweating, in spite of the extreme cold. Then came the pushups. After 50, Jay was sure he could not do any more from the way his muscles were screaming. A screaming JUO convinced him otherwise with a flow of high pitched invective. Maybe the flesh could still be driven even if it was weak and even if the spirit was highly unwilling, That is if the screaming was loud and content scary.

Then out of the mist, came yet another duty officer. He stood there for a minute, unnoticed by the appointments. The I termers saw the officer and froze and the BCA jumped when he realised there was an officer behind him. The I termers were still traumatized by the betrayal of the first duty officer Capt. Jeevan Singh. They did not have very high hopes from this one.

The officer in question was a Capt. Ali Shah, a well known platoon DS or directing staff. His claim to fame was being soft-spoken and merciful—a rare trait among platoon DSs. He was also known for wreaking havoc amongst those who crossed him...after the third strike.

Capt. Shah had been sent specially by the battalion commander or CO (commanding officer) to curb the blood lust of the III termer appointments. The CO, in his long experience in service, had seen the old saw in action that new shoes tend to bite. In this case it was the new cadet appointments who were eager to prove themselves. They were especially keen on testing out their newly-acquired powers. He had had misgivings about the efficacy of Capt. Jeevan Singh in preventing the oppression of the I termers. He had been right.

Capt. Shah asked in his characteristically gentle tone, "What the fuck is happening here?!"

The BCA gulped. The only thing worse than a known devil was an unpredictable one. BCA Binny Joseph had not been able to fathom what made Capt. Shah tick. But he did know that Capt. Shah had lost a brother in the NDA to a punishment parade and was quite protective of the I termers. He was also known to be quite opposed to unauthorised 'punishment parades' while supporting organised parades designed to 'toughen' up cadets.

"Sir, we were told by the duty officer to warm them up..." and the BCA's voice trailed off.

The I termers perked up. The exchange was offering them their first ray of hope in the whole evening...and then it came.

"I want all the appointments here in two minutes...and tell these bloody I termers to break off." said Capt. Shah and ominously added, "Your bloody time starts now!"

There was restrained chaos as the BCA and his minions fell over themselves to comply with Capt. Shah's orders...within the time stipulated.

A few minutes later the I termers were in their rooms shivering quietly and hoping the day was over. They heard the third termers getting what was described as and dreaded throughout the Indian Army,' The bollocking moral lecture' or just ML. It's an extreme version of a scolding. It leaves one with the feeling of having being kicked in the crotch while being mentally crushed. The numbness sets in only after the first half hour but the pain returns after one hour of the 'ML'.

Jay quickly changed out of his dew-soaked clothes and was about to settle into his bed when he heard a knock. He felt a deep foreboding as he opened the door. There was a II termer waiting for him outside. But Jay did not recognize him and thus made the mistake of asking curtly, "Yeah what is it?"

"Shut the fuck up, put on your PT dress and stand next to your bed. You will not lie down till you are so instructed...Now shut the bloody door and do what you have been told." The instructions were issued coldly, without any perceived need for explanation.

As Jay peeked out he saw other II termers, in their coats and ties, passing similar orders to other I termers. Something momentous and traditional was up and Jay did not want to come in the way of any such tradition. He hurried to carry out the orders and put on the wet dress once more. He later learnt that it was Academy tradition that if the seniors are being screwed, then all the juniors had to wait up and share the suffering in part and spirit.

After some time he decided to take a risk and wrap a blanket around himself. He dozed off and was woken again by the sound of frantic knocking. It was Anant Kumar, looking very disappointed.

He gave the message he had been tasked with. "The duty officer has left the III termers. You can go to sleep now. Muster fall in at five!"

Day five of Jay's dream was officially over...it felt very much like a bad wet dream considering the amount of moisture he had soaked up. But the end game, when he would be a cash-rich young bachelor, with lots of adoring women told his psyche to stick it out. This was where 'the flesh was unwilling and the spirit was weak, but the fantasy strong enough to keep both going'!

Anant's fantasy of an office was coming to an end. He was beginning to accept the fact that the Indian Army was not really for him. He had hoped to become a government servant and help the country. But here he was worse than a servant, and inside the Academy, no country could help him.

NOTE

1. 'De tab' IMA slang, it means being stripped of rank or appointment, a euphemism for removing the shoulder tab or boards the cadet appointments wore to signify various ranks or appointments.

Chapter 9

END OF DAYS !

He had been in the IMA for about two weeks now and had actually lost count of the days. The first week had been a blur of familiarization, issuing of clothes, précis and the cursed bicycle. The bicycles rarely worked and were almost always punctured. The academy was very insistent about movement only in squads. That meant the bike-less unfortunates had to jog along with their more fortunate mounted coursemates. And then there were the drill parades and punishments where the bikes came in very handy for an inventive senior.

The Academy was all that Jay had dreamed about, with a salubrious environment and martial atmosphere steeped in history. If only something could be done about those who made it a nightmare. He had a nagging feeling that his dad, Mehta and the other cadet idols had not told him the whole truth about their 'wonderful' days at the Academy. He understood now that junior days were to be forgotten. What he could not understand or forget was the oft repeated Academy saying, 'When rape is inevitable, lie back and try to enjoy it!' It was difficult to embrace that concept when you got raped every few hours.

The cycles were versatile torture and punishment tools, rather than modes of conveyance. The cycles, due to the perverse nature of man and military mind, had turned tools of convenience into convenient millstones. They could be hung around an offender's neck and later, ridden off...in squads and not conclusively into the sunset. The use of the cycle could be denied, even turned into a burden and made life even more uncomfortable. Truly, the cycle was made for the IMA and transcended the expectations of its manufacturers. If they had used cycles in Guantanamo Bay camp, they would have finished off their business long back. As torture techniques go less messy than waterboarding...

The days seemed to be divided into PT time, parade, class time and fall in time. The fall in's seemed to take forever. Jay could not find it in his heart to forgive the plethora of appointments for their sheer ineptitude born out of inexperience. This learning curve had a sharp edge and many pricks...pun intended.

The weather did not help the situation when they had to stand for hours, while the parade state was compiled or some appointment felt like ranting. It was cold...very cold, and the minds of most free-thinking mammals, especially the cadets, were numbed. The masochism of those who had joined out of bloodlust and quest for martial glory was at an all-time low. The motivations for joining the army were being actively rethought.

Jay had set the alarm at 4 AM and it rang, loud and true. It did not, however, feel true in his head or heart, both of which told him he needed more sleep. He felt as if he had a hangover and the cold was insidious, it was all-pervading. Jay had spent a large part of his life in the North and was thus used to cold weather. Some of the others, like Anant and Ramulu, were traumatized.

The other aspect which was the biggest pain was being at the beck and call of the III and, to some extent, II termers. The IMA motto seemed to be 'Service before Self'. This aspect was enforced well. The I termers seemed to be engaged in service of others with very little time for themselves.

Jay opened the door and saw the coast was clear of predators. He rushed out to the bathrooms and was stopped in his tracks by a number of issues. Issue number one was that he had forgotten his dressing gown, a felony of sorts in that place. The other was gentleman cadet Lal Dev, an ex-NDA II termer and practised predator. He had issues regarding the conduct of toilet without stimulation. He brought these issues to Jay's notice succinctly, "Fucker, where's your fucking gown, take this mug and get me tea! Make it fast I'm feeling crappish."

It was too much information and not too subtle blackmail. Jay preferred to think of it as a life saving mission, only he was not sure whose life. There were also other things Jay was not sure about so he asked (very politely). "Sir, where is the *chai* available?!"

He got the abrupt reply, "Fucking find out, fucker where the fuck the *chai* is!" Having ensured the 'f' word had redundancy GC Lal Dev also nicknamed 'Red God' (a literal translation of his name) exercised his divine prerogative and slammed his cabin door.

Jay looked around wondering where the tea was. There was no one reliable he could find out from with getting burdened with additional chores.

Only the I termers and the II termers were up and about in the semi darkness of 5 AM. The III termers would get up a little later. It was the juniors' dubious privilege to clear the bathroom first. But before that, Jay had to finish bringing tea and run other errands like getting a belt, rifle sling or some such item from somebody for somebody, both bodies usually senior.

After some hunting, Jay found the tea thermos that the mess staff had conveniently put in the centre of the road junction. There was a long line of shivering I termers, anxious to get back to their masters, before they got impatient or worse...were overtaken by the call of nature. A displeased senior was never good news in the Army. Jay, numbly, waited his turn and pondered his circumstances and life under a new master.

The master in question was not GC Lal Dev, but another III termer named Rathore. Rathore had gotten in a few days ago after the term break with the rest of the course. He was not a tabbed[1] appointment but just a corporal. He was thus powerful and outranked GC Lal Dev. He also coveted Jay's services...so to speak.

Jay had first noticed Lal Dev about three days after joining the academy. GC Lal Dev had knocked very diffidently and asked very respectfully (even called him 'sir'), if he could borrow a bucket. Jay had breezily told him to take it. The exact words, which Jay had cause to remember later were, "*Le jaa yaar*!"

The next day he noticed Lal Dev wearing an NDA blazer, which meant he had to be a II termer and he realised that he had been rude to a senior. The next day Lal Dev came back to request for the bucket again.

Jay had sprung up from his bed keeping with Academy etiquette, and said, "Yessir! Sure sir."

GC Lal Dev seemed to 'grow', or maybe the term that would describe the phenomenon more aptly was 'puffed up'. He was outraged and seriously embarrassed at the indignity of having shown servility to a social and professional inferior.

Lal Dev called down retribution on the spot by making Jay do 70 pushups. He completed the subjugation by ordering Jay to fetch hot water for a bath from the *hammam*. The final humiliating nail was Jay was forced to use the same blessed bucket Lal Dev had come to borrow, before Jay had given the game away. Over the next few days Jay had become a flunkey expected to wait hand and foot on Lal Dev.

And then GC Rathore had landed on the scene, he was a III termer, a corporal and senior to GC Lal Dev. He did not like triangular relations, professional or social. As far as GC Corporal Rathore was concerned, in a master-slave relationship, three was a crowd. Jay was required by Rathore in order to make his final term in IMA comfortable; Lal Dev was just a minor irritant. Jay would have not have agreed with that assessment of Lal Dev, he was a major pain in the neck.

Jay wished he was just a 'one man dog' with a clearly recognizable master. But GC 'Red God' had been dropping not too subtle hints. The most explicit of these had been the one about GC Rathore passing out in six months. Thus, leaving Jay's 'friend always in need', GC Lal Dev in charge. Jay began to understand, all too clearly, the meaning of the word 'chattel'.

Jay suddenly woke up and realized he had fallen asleep while standing in line for tea. When his turn came, he took the tea and rushed to Lal Dev's cabin but was neatly intercepted by Rathore. "Thanks for the tea!" was all he said. Jay relinquished the tea to Rathore and after briefly considering going back to get some more discarded the idea. Jay was horrified at the consequences of being caught in a conundrum of military hierarchy but he was also getting late.

There was only one thing he could do now. Go and get ready for the drill parade before he offended anymore of the demons waiting to trip up I termers. He dashed off to the bathrooms and spent the next ten minutes avoiding Lal Dev.

The critical time passed and all the I, II and third termers were in a mad rush to go get their rifles from the *kote* (armoury) and rush off for drill. All this had to be done in even numbered squads of four or more or one had to run with his cycle, a physically excruciating and tiresome exercise closely monitored by zealous drill instructors. No one had any time to waste; though Jay was sure Lal Dev would 'sort him out' later.

It came that very evening. Right before dinner, he saw a note on his door that was to the point-'triple pack *bajri* (gravel) order, 1830 hours in front of my cabin.' Jay did not need to be told whose cabin; it was Lal Dev's. He did have some doubts if he was really expected to come weighed down by three packs filled with *bajri.* That would be around 60 kilos, almost his body weight. Seemed kind of excessive punishment for a cup of tea, but then as Jay admitted to himself, drill without satisfactory bowel movement could be quite discomfiting.

Still Jay could not resist the idea of letting the master Rathore know that his slave was being misused by another overseer. So Jay staggered under the weight of the three pack 08s[2] to Rathore's room, knocked and waited with bated breath. Corporal Rathore had Lal Dev kicked out of his cabin shifted in next to Jay. It was not out of affection, but a mere practical step to ensure proximity to service providers.

"Who the hell is it?" came Rathore's voice.

He had been woken from his nap. Jay hoped he would be mad and maybe...just maybe, Rathore would excuse him from being Lal Dev's amusement for the evening...or so Jay feverently hoped. It did not happen.

"Sir, I got a note to report to you in triple *bajri* order..."

There was a moment's pause and then Rathore broke the suspense, "Yeah, report to Lal."

Jay was disappointed, anxious and intrigued. The last emotion was because it seemed Rathore was aware of Lal Dev's invitation. So young Jay, weighed down by disappointment and gravel, trudged the ten long steps to GC Lal Dev's room.

He knocked and waited with heavy heart and packs. After five very long minutes, Lal Dev opened the door and looked long and hard at Jay. Jay looked back and was a bit surprised at what he saw. Lal Dev was wearing dungarees and looking ready for combat, though his expression held a tinge of fatigue and anguish.

He broke the suspense, "Fucker... why didn't you get me the tea..." This was stated with what sounded like a twinge of sadness

Before Jay could answer Lal Dev continued, "...Because of you I got late, and now...take off your pack and help me put it on...!"

Jay did as ordered, his heart beating with relief. Lal Dev put on the packs with Jay's assistance and then staggered off towards the BCA's room. Jay went around the back and knocked on the BCA's understudy's window.

This was his course mate, a I termer like him named Bhasin. Fate had been kind to GC Bhasin; he was the BCA's understudy. His job content, irrespective of rumours, was much like that of a private secretary-cum-butler. The fact that Mr. Bhasin was somewhat effeminate and downright bitchy at times greatly supported the various theories floated by his coursemates. The theories were, of course, all to do with the true nature and extent of services provided to the BCA. Bhasin had been spared all 'fuck fall-ins', in fact ALL fall-ins.

An apt description of his duties would have been, man Friday, Saturday, even Sunday; in fact Man 24/7. There were other less flattering terms to describe Bhasin as he strove tirelessly to keep his master and 'overstudy', the BCA, happy. He was a born personal staff officer-type. He was bent on becoming the quintessential, though historically reviled, PA or personal assistant.

That the BCA himself had once come to rescue Bhasin from a particularly pesky 'lancee' or lance corporal (the only II term appointment) had greatly added to Bhasin's aura of privilege. There were many who wished to take away that immunity, most of those being Bhasin's own coursemates.

Jay sneaked around the back and knocked furtively. Fraternizing with Bhasin was not exactly good for one's standing in Academy circles. GC Bhasin opened the door with a haughty look. The look was very much similar to that a hooker, who has now become the local don's moll, gives her erstwhile lowlife companions. But Bhasin had to be tolerated. Jay was interested in finding out the reason for Lal Dev's date with the infamously strict BCA. Relief at being spared a horrible fate can often be supplanted by curiosity.

Bhasin acted displeased at the intrusion and managed to convey to Jay how busy he was compiling the parade state and other vital company matters. But Jay knew Bhasin was starved of human contact. The other seniors with whom Bhasin normally cavorted, treated him like some pet monkey.

After all it was better to be a 'servant in heaven than a king in hell'...a very popular line of thought in IMA, if not in heaven. But the fallout of this meant that he was deprived of an audience, being unpopular among his coursemates. Bhasin was actually in an expansive mood and keen on sharing anecdotes of his importance in the system. Jay's arrival thus came as a godsend on that lonely evening...one of many.

He willingly told Jay how the dragon Lal Dev had come to be vanquished by the valiant BCA. The story began with Jay's non-delivery of tea which held laxative properties for many, and especially for GC Lal Dev. The tea not being forthcoming, he had set off in a rage to get the tea himself. He had found the flask empty and thereafter wasted time looking for tea from another platoon, unsuccessfully. Things had gone really south for dear Lal Dev after that. He had gotten late and could not find a squad. He decided to take a risk and sped away alone through a shortcut to the drill square.

That was when Lal Dev had made his second mistake. His cycle had a flat tire and so out of desperation; he had picked up the first cycle he found. This happened to be a III termer's cycle with said III termer being the BCA's chum. The matter was brought to the attention of the BCA who was very, very 'wroth' to put it mildly. But that came later.

GC Lal Dev had been barrelling down a lesser-used path to the drill square. He did not have a squad and only appointments were permitted to ride cycles without a squad.

This was in the forlorn hope of avoiding drill instructors, duty officers and other lesser tigers like JUOs and BUOs. It was foggy and still dark at six in the chilly winter morning. But one often meets his fate on the road taken to avoid it. And so it came to pass that GC Lal Dev ran into the BUO, who was knocked off his cycle, but still managed to recognize Lal Dev as he fled.

It was going to be a long and torturous night for Lal Dev. Jay could see him standing there, legs quivering under the weight of the triple *bajri* pack and two cycles. He was steaming and sweating and Jay almost felt sorry for him. It was just a matter of time when the roles would be reversed. But the satisfaction of seeing the tormentor tormented was very pleasant indeed.

Jay silently praised the Almighty for this shining example of divine justice. He went back to his room a happy man. The next few hours, and dinner, went off uneventfully. He had just about settled in for the night when he heard the announcement for a special I termer fall in the next day.

The II termers rallied in support of Lal Dev and had sought the CSM's blessing to sort out the Ist termers. It was on a Saturday afternoon and a rare half day off. Thus all II termers were available to supervise with company officers and other instructors notably absent. It was also raining which added a kind of carnival aspect to the whole scene. The I termers shammed it out with steely resolve. Their reactions were slow, slow enough to infuriate but not enough to escalate the punishment. The II termers after a couple of hours lost interest and started trying out variations just for the sake of formality.

The I termers were barely recognizable in their swimming trunks covered in mud and grass. GC Chib had been given a 'balli' (log used for PT exercises) and was standing like a dirty mini Hercules in his red IMA swimming trunks and mud embellishments. His parents were due to visit him. Chib had forgotten, and worse he had forgotten to mention the same to the II termers that he was expecting guests.

A 'fuck fall in' could always be convened later, but the image of the IMA was paramount. Civilians and other vulnerable relatives like mothers were not supposed to be exposed to the brutal side of military grooming. Unfortunately due to GC Chib's memory lapse the 'faux pas' happened.

The senior lancee overseeing the punishment orgy noticed a middle aged couple approaching the PT ground anxiously. They were Chib's parents. Chib's saw them and moaned something that sounded like "ohhh shiit". The II near Chib's heard that and realised it was someone known to Chib's and rushed to the II termer top dog. The top dog switched on his charm and replaced his grim master look (as he fancied it) and replaced it with a polite smile 'faux' enough to match the 'faux pas'.

He asked Chib's father, "May I help you Sir?"

Chib's father saw what looked like a scene from 'Apocalypse Now'. Mrs Chib's having that womanly intuition suspected her delicate darling (GC Chib's would have disagreed with that description) was nearby.

Mr Chib's asked, "We're looking for a GC Chib...I'm told he is here doing training..." This last he said without much conviction.

Mrs Chib's asked nervously, "Uhh... inspector *sahib*...what's going on?"

The senior cadet gritted his teeth at being referred to as inspector. But truth be told he looked like some inspector on a raid. It did not help he was wearing some uniform Chib's folks had never seen before except maybe on some guard or police type. The fact that they were nervous about the whole scenario did not help their deductive process.

Chib's senior's nervousness was justified. He had imagined his son had gone to become an army officer, instead there seemed to be some kind of extremely debauched rain festival going on. There also seemed to be a heavy involvement of mud and acrobatics in the proceeding. Strangely no one seemed to be smiling or having fun. The II termer in charge realised things could get out of hand anytime called Lancee Pandey over.

He said, "Sir, maa'm, this gentleman cadet will escort you to the waiting room while we locate your son."

Mrs Chib's asked again hopefully, "My son isn't here?"

L/Cpl Pandey was acutely aware of GC Chib's location some twenty five yards behind him. It was just five minutes back that he had advised Chib's of an imminent kick if he did not hold the *balli* straight up. In fact the exact words had been "hold that fucking *balli* up you weak bloody pussy or I'll give you such a kick in that arse you're sticking out like a tired whore that my boot will tickle your tonsils!" It was a different matter that both had never laid their eyes on any whore tired or otherwise.

"No ma'am he is busy studying at the library. I will send somebody to get him while you wait in the mess ante room." Said lancee Pandey brightly.

Mrs Chib's looked horrified, "That looks like my...son!" She said pointing at the mud covered caricature of a human who only a mother could have recognised.

Lancee Pandey tried his most charming smile and said, these are just some boys who have been naughty, and are now practising for the..." Words escaped him at this juncture.

GC L/Cpl Pandey's smile worked quite often on his own mother, but it did not seem to be working on GC Chib's mother. Mrs Chib's was still looking horrified and pointing at her son. GC Chib was torn between fear of eternal ridicule and the desire to run to his mother and hide behind her maybe even make her spank those bloody II termers. Actually the entire lot of I termers wished somebody would spank those pesky II termers. The feeling was mutual at that point.

It was about then Mr Chib's recognised his son. Fathers don't have the same instincts as mothers, but luckily quicker reactions. He quickly grasped the situation and said, "Come on *ji,* this is how they do training in the military. I'm sure our son will be coming from the library. Come on you... lead us there."

Mr Chib's said all this very fast and dragged his wife off before she could positively identify her progeny. Hell hath no fury like a mother... well any mother with sorely abused offspring.

Pandey led them to the anteroom with great relief. Two II termers rushed up to GC Chib and dragged him to the bathroom hosed him down and stuffed him into his muftis. The whole action from the time they laid hands on him till they dumped him at the anteroom entrance, took exactly six minutes. GC Chib was still towelling his head when he was given his last briefing before being shoved into the anteroom, "You are enjoying yourself at the academy thoroughly...and you were reading about Montgomery...understood?!"

Chib handed Lancee Pandey the towel and nodded. He wanted to ask who Montgomery was, then smiled painfully, checked his reflection in the mirror and walked into the anteroom...smile frozen. Pandey mopped his brow with the towel and waited till he heard Chib's mother squeal. Then satisfied that the family reunion was back on track walked back to the PT ground to break off the other I termers. He had had enough.

NOTES

1. Tabbed appointments were the ones who wore shoulder boards and were considered a cut above those who wore stripes.
2. Pack 08 so called that because they were backpacks introduced in 1908 still used till the late 1990s as part of field kit. Very hardy and extremely uncomfortable.

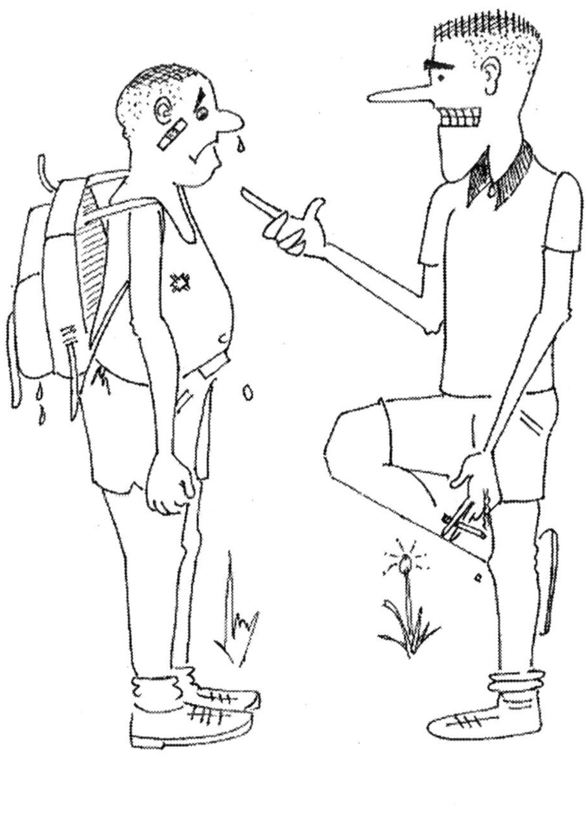

Chapter 10

THE PECKING ORDER

The phenomenon of PT after dinner and in the freezing cold, was something Jay and the I termers could not quite come to terms with. The matter had been discussed at length, but no one had discovered a way out or courage. Many aggressive declarations were made, many spirited oaths taken, but nobody had refused to come for the night PT. Every course has a hero. Heroes tend to survive and become famous but those around them tend to do poorly. Heroes are not very popular in the academy. Jay's course had one too and unfortunately, in the same company.

And so it came to pass, after weeks of waiting for his office and suffering indignities at the hands of '*mofussil*' (small town) barbarians' one worm began to turn, even squirm.

Finally, that fearless scion of a freedom fighter, Gentleman Cadet Anant Kumar had gone to the company commander. This was Major Ajay Singh, a highly decorated soldier with once great prospects, currently a prospective alcoholic. The gentleman was recently divorced, presently always irritable. But GC Anant did not know that and he did not care. All he wanted to do was convey his displeasure at the unchecked ragging that was taking a toll of the I termers' morale.

Maj.Ajay Singh was amused to see a I termer hanging around the battalion office. It was such a beautiful day too. As luck would have it, the CSM and CUO were also there to take orders, while he sunned himself outside his office on an easy chair. He called Anant over and asked if he could be of assistance. Maj. Ajay was a thorough gentleman, at least outwardly, and one who never raised his voice. It was this courtly attitude that fooled many people. GC Anant was also fooled and felt here was a man who would save them from the terrible injustice.

"Sir, I have come to complain about the ragging that is taking place in your company!" GC Anant Kumar did not beat about the bush.

"Really?!! I think you mean our company..." said Maj. Ajay with a sarcastic smile and turned to the CSM and CUO, "You gentlemen are not looking after our I termers it would seem..."

The CSM and the CUO simpered nervously. They knew fully well what the Maj. Ajay Singh was capable of and how unpredictable his temperament. If GC Anant had been a mouse and if cats could smile, he might have recognized the smile. It was one a sporting cat would give a mouse with terminal prospects.

But GC Anant carried on blithely, "Sir they are ragging us every night and making us do menial jobs!"

"Ragging? Oh dear...Which platoon is this poor harassed boy from?" asked Maj. Ajay of the cadet appointments who were now looking positively worried.

"Sir, second platoon, Capt. Jeevan Singh's...sir..." they nervously replied. They knew from experience that the more polite and courtly '*ajgar* (python) Ajay' got, the worse the future was for the audience.

Maj. Ajay picked up the intercom and called Capt. Jeevan over to the company commander's office.

GC Anant was satisfied that things were moving, and that finally, he would be treated as an officer should be. He had given up hopes for an office some days ago. The CSM and CUO were not sure what the company commander's intentions were, but it seemed likely that someone was about to get squeezed. The python's coils were tightening...

The appointments tried to cover their asses, "Sir this chap has never come to us with his problems..."

Maj. Ajay waved them off, "I don't give a fuck! But he has come here...no? To me... Obviously you gentlemen can't handle simple things and can't teach the I termers about life in the glorious IMA..."

GC Anant was still blithely unaware of the undercurrent that was sending shivers up the CSM and CUO's spines. He stood there happy that stern action was being taken based on his complaints.

Capt. Jeevan came in, sized up the situation accurately and asked, "What does this freedom fighter want?"

Anant was elated he was being referred to as a freedom fighter just like his grandfather. He still did not know that this was a derogatory term. It has been in use since the days of the 'Raj', commonly used to refer to someone who liked to buck the establishment. Almost every new term had a few such freedom fighters who popped up only to be squashed. They needed to be put in their place. Freedom fighters were dangerous. They encouraged questioning of orders and needed easy answers to difficult questions. And of course they wanted change. And as any military person in any academy or dictatorship knows, change has always led to the gradual breakdown of the system.

Maj. Ajay and Capt. Jeevan looked at each other. They were acutely aware of the above mentioned train of thought. Their thoughts were in harmony and goals, in unison. Sort out the I termers and make sure the third termers understood their responsibilities. Make officers of 'em in the end and soldiers along the way. It was in the 'along the way' process, where the instructors came in.

"Jeevan, this young man here feels that he and his mates are being 'ragged'... He was cribbing about some night PT and menial jobs and some such bull crap. I would like you to check it out this evening after dinner. See if you can convince them. Ragging huh? Have you ever heard of it in the Army?! You are the duty officer also, aren't you?"

Capt. Jeevan got the message and nodded. He understood perfectly what had to be done. He knew how to put down potential dissatisfaction. He turned his cold expressionless eyes on Anant, then the CSM and lastly the CUO. They all felt a nasty tingle of anticipation. Only Anant stood there with a smug look.

"Don't worry sir I will remove all doubts... " Jeevan then turned to the appointments, "CSM fall in after dinner. I will take the report."

"Maybe you should get the whole battalion fallen in, then you can explain to them all the difference between ragging and toughening up..." suggested Maj. Ajay

The CSM and CUO nodded their heads silently in approval and relief. They were glad the company commander had decided not to include them in the education process. The looks they gave Anant Kumar would have stopped even his illustrious grandfather in his 'civil disobedience'[1] tracks.

Anant conveyed the happenings to the rest of the I termers. Most of them could not read the writing on the wall. They were highly expectant of being freed from the bother of night PT and unpaid servitude. But there were others who had their misgivings. They all prayed that things wouldn't get worse. A general air of unease prevailed amongst the I termers of Bhagat battalion.

Jay had almost forgotten the possibility of an interesting night after the Lal Dev episode. But the announcement brought back his feelings of dread. He did not share the optimism of some of his coursemates that Anant Kumar's whistle-blowing could change things. He did know that complaining about your senior never ended well for the junior in the Army. He had a bad feeling about this night.

Jay picked out his PT dress. He looked at the it sadly; it was equally sad-looking. It was browner than white, especially the jersey. He had four sets, two issued and two his own. But they had not been able to keep up with the pace of PT and punishments.

Most of these exertions seemed to involve rolling or crawling on dew-drenched grass or dusty roads. Jay wondered which genius had decreed the wearing of shorts in winters and white clothes in any military academy. His knees were terribly exposed and the wind seemed to blow right up and chill his family jewels.

The I termers were assembled company-wise and then herded by the lance corporals. The 'lancees', were the only appointments or ranks II termers could have and they were usually a harassed lot. Some of them, at least the direct entry types, were quite sympathetic to the I termers' plight. The buck stopped with the lancees who did most of the lower level coordination and management. But they had no special privileges or immunity from official punishment as the other appointments had.

Their shepherd was lance corporal Rajesh Pandey, a very decent sort—as lance corporals go. He tried to be as humane as possible and kept the brutality levels down when he had to carry out punishments. He was a Bihari from Delhi and was rumoured to have a very beautiful girlfriend. The I termers loved him. The rest of the III term appointments and the ex-NDA II termers thought he was too soft.

They reached the 'execution square', the area in front of the battalion mess. Their host for the evening's entertainment was the BCA. The other company commanders had decided to contribute their own I termers to the evening's entertainment. This was now a battalion level 'fuck' fall-in. There was a message that was to be broadcast widely.

The BCA was tense because the parade state was not ready. He was badgering the CSMs, who were in turn yelling at the JUOs, who were in turn leaning on the lancees. It was a typical military scene of controlled chaos, repressed feelings and barely controlled rage. Rage in an academy full of testosterone-driven young men, some of whom are in charge of others, is bad news. It was very bad news for the I termers, some of whom were still in a state of denial.

"Maybe they're just going to give us a pep talk or make some announcements..." whispered Chib to Jay.

The I termers were woefully outnumbered by the sheer volume of the other terms. Their combined menace was a heavy burden that crushed all thoughts of disobedience reprieve or rebellion among the I termers.

Jay looked around at his coursemates. There were karate champions, body builders and heroes from lesser towns and famous colleges from all over of the country. Yet they all stood meekly, awaiting their fate, their eyes and spirit downcast. The need of the hour was not to attract attention. Everyone seemed to shrink as they tried to avoid the gaze of the JUOs who were striding up and down, trying to find someone to make an example of.

The BCA, who had been simmering for some time, finally lost his cool. The catalyst for this calamitous happening was the rebellious II termers, especially the ex-NDAs. They had gone from being lords in NDA to serfs in IMA. Most of them had not yet come to terms with it and tried to show their dissatisfaction with the new order through a sullen and slow reaction to any orders given.

In this case the orders were to get fallen in so that the report could be compiled. This was not happening fast enough. There were just about 40 I termers who posed no problem; they were standing meekly, totally subjugated. Anant was just a mere insect that would fly away soon. They had seen this sort before, the IMA was not for them...it was just a matter of time.

The III termers were lolling around the anteroom, playing billiards or just watching the fun. The BCA did not have much control over the III termers as they were his coursemates, though technically he could cause them serious trouble. That would have resulted in the BCA's social boycott, there being no higher sin than 'fingering' a coursemate. But the II termers were perfect targets. They were neither senior nor very junior. They were perfect for establishing ascendancy

The BCA's main job was to maintain discipline. The more feared he was, the more he could impress the instructors. That would translate into a better standing in the merit list, giving him a better opportunity to get his choice of regiment or corps.

The only thing standing between the BCA and his dreams were these II termers. Capt. Jeevan was due any time and the report wasn't even halfway compiled. There were no signs of order or control. The BUO was of no help—he was standing behind a pillar, smoking a cigarette.

The BCA called over one of his favourite henchmen, JUO Akshay Kumar. That worthy was also known as 'Keen Kumar' by his coursemates because of his frequent displays of extreme keenness, especially when the officers were watching. He was however more popularly known as '*Kutta* Kumar' because of his doglike demeanour. He was like a happy puppy in front of seniors and a real SOB otherwise.

BCA Binny Joseph muttered a few instructions to Akshay Kumar and unleashed him. JUO Akshay deputised two likeminded sergeants as fellow inquisitors and advanced swiftly on the gang of mutinous II termers. The II termers knew there would be consequences but considered themselves tough enough to handle whatever the III termers could dish out. They were also under the wrong impression that the night's victims were only the I termers and that they would be exempt.Their doubts were firmly dispelled in the first few minutes. Akshay and his two sidekicks positioned themselves strategically.

"Get into pairs, fuckers!" said JUO Akshay Kumar. The action had begun, albeit a few minutes before Jeevan would have formally inaugurated it.

As the I termers were learning, whenever a senior, in the mood for a little entertainment, looks far into the horizon it does not mean he is admiring nature or cosmic karma and his future. He is merely looking for a landmark to serve as a halfway point, around which the tools of his entertainment could be made to race around. Jay was just dazzled by the way the scope of the evening had widened. He just hoped whatever the size of the party, the I termers should not be considered dessert and had much later in the night...again

The seniors requested the intended victims to get into pairs. It was not for sparring or dancing; it only meant they were avoiding the inconvenience of getting triple '*bajri*' packs. Using pairs and their body weight also meant a larger number of fit and clever young men could be gainfully engaged in one spot. These were management tips, learnt the hard way through personal experience and administered through word of foul mouths.

The II termers were suitably impressed by the BCA's initiative. But ego demanded they show some defiance, some assertion of authority. After all the I termers were watching. And they the II termers would be masters in, but a few months. They did not act with JUO Akshay's desired speed or as he put it, "The reactions are too fucking slow!"

"OK! All II termers round the mess and the first two pairs will be left off. !" announced Keen Kumar.

The II termers realized that things were getting serious. There was a mad scramble as they all raced around the mess. The I termers watched in awed terror as they waited their turn with the executioner.

And then the chop came-one of the sergeants noticed the I termers and screamed, "Fuckers! Your pops are being screwed and you are LOOKING! Look at the fucking sky!!"

The I termers looked up at the sky so fast they could have dislocated their necks. But that was not enough. BCA Binny gave a subtle nod to JUO Keen Kumar and thus signalled the beginning of the first official 'night fuck fall in' of the term for the I termers.

"Get down on your hands bastards... I mean gentlemen!" requested Akshay Kumar in his signature style. A couple of his 'three striper' cronies laughed dutifully. The irony was a bit humorous; it being against the rules, the written ones at least, to verbally abuse gentlemen cadets.

The physical abuse part was covered by the clauses of training and punishment as training. Rolling, crawling, etc. was part of the fieldcraft manual. Everything else was covered in the PT handbook of freehand training and musketry.

The first few II termers came around the mess block and the winners were neatly intercepted and made to get on their hands 'down' and commence pushups pending further decisions. The rest were neatly corralled and assembled in pairs.

The I termers were a bit disappointed by this flagrant breach of contract, after all the winners had been promised a better time. But as events unfolded, they realised and would remember that being on their 'hands down' was infinitely better. At least better than what JUO Akshay, the Keen Kumar, had in mind.

There was another group that excited the envy of all terms. They were the foreign cadets. It seemed the average Indian's partiality for things 'imported' extended to cadets from other countries like Iraq, Bhutan, Sri Lanka and Mozambique to name a few. Most did not attend any fall ins. They had come for this one, because they had been tactfully requested to be present for the duty officer's routine check.

Keeping the foreign GCs hanging around was about the maximum the cadet appointments could do to 'finger' them. It was very frustrating for all concerned, starting from the DS body consisting of officers, JCOs and NCOs, right down to a lowly II term lancee. The whole bedrock of training in Indian military academies was about inculcating total obedience to orders, no matter what..

The I termers felt the desired level of frustration, and that increased as they saw the foreign cadets loll about without fear or care. There were some, like the Bhutanese, who having gone through three years of NDA training with their Indian coursemates, felt sufficiently motivated to join them in their time of adversity.

But that noble sentiment was mostly lost on lesser imported cadets who were of a normal psychological bent. Why, at that point of time there were many Indians who would have gladly sold their souls to be back in their beds.

Jay for one wished he could be back in his bed, along with the rest of his coursemates; actually so did everyone all over the Academy. So did Chib who was cursing life, his father and all factors responsible for his current position. What the victims did not know was that the same scene was being duplicated in other corners of the Academy, in varying degrees of harshness.

The II termers were getting warmed up. Judging by the sweat streaming down the toned and relatively trained II term bodies in 'minus' temperature, 'getting warmed up' was an understatement. JUO Akshay Kumar was in his element.

The one great 'saw' JUO Akshay Kumar lived by was-"sticks and stones may break my bones, but names will never hurt me!" He said as much in whatever English he could muster. The collective invective the II termers had accumulated would in itself have had considerable weight to crush anyone.

The front row of II termers were carrying their comrades, sitting upright, on their shoulders. The difference in posture made a lot of difference, literally and figuratively. The weight and physics of a person sitting upright on one's shoulders made many muscles come into play. All those muscles were now strained to their utmost limit.

Two shifts had already changed, that is the lot that had been sitting on top had already taken their turn below after a count of twenty...twice. They were breathing heavily and were beyond egos or any macho pride. These were II termers and semi-trained soldiers, they knew the drill and survival was the key. 'He, who took the shit manfully, lived to dish it out to others.'

The BCA's ego was somewhat mollified by the sight of the heaving and sweating II termers. He concentrated on completing his parade state. There were still some cadets unaccounted for and the strength was not tallying but then it rarely did in times of stress.

The stress levels shot up with the appearance of Capt. Jeevan, in combat dress, riding his bike. He cut a dashing figure and the cadets at once admired and loathed him. JUO Akshay Kumar idolized Capt Jeevan. The rumour was 'Akshay wanted to grow up and become an asshole just like Jeevan'.

The third termers scrambled, the BCA fumbled with his parade state and his understudy Bhasin melted into the shadows. The BUO tossed his cigarette, jumped out from behind his favourite pillar and started looking busy and in control of the situation.

"Battalion *savdhan*!!!" the BUO yelled as the III scrambled from the anteroom to get into attention.

The BUO had been chosen for the appointment mostly because of his good looks, sportsmanship and impressive voice. The brain part was handled by the BCA who was at that point, feeling rather numb.

Capt. Jeevan took his time parking his bike. Like any good inquisitor, Jeevan knew that anticipation of pain increased terror. It furthered his reputation as a bastard and that gave him a further high.

He surveyed the scene and was satisfied with the level of suffering he saw. The I termers were still on their hands down in push-up position. The II termers' heart rates were still quite high and the third termers were avoiding his gaze. He turned to the BCA who rattled out the parade state.

"Is everyone here...?" Jeevan asked casually.

'Yessir, only the category types and some others I have detailed..." said the BCA with as straight a face as his nervousness could manage.

"How many I termers and third termers are absent?!" The II termers like anyone who has been a middle child knows, got no leeway, they would all be there.

The BCA was in a quandary, he had excused a couple of cronies and of course two first termers, one of whom was his understudy and the other his 'school type'. Falsifying the parade state was a serious crime and could mean relegation.

The BCA decided to give it up and save his ass, "My understudy and four cadets who are category...sir..."

"Get them all out the fuck here now!" said Jeevan, and walked off to his bike.

The BCA's heart sank. It was obvious that Jeevan was going to check the cabins and drag out anyone he found. He turned around and saw with relief, his understudy Bhasin hiding in the shadows. Bhasin's voyeuristic eagerness to witness a fine 'fuck fall in' might have saved his worthless skin.

BCA Binny was to the point, "You bloody well run and warn all the cadets in their rooms, Capt Jeevan is coming... if you get caught say I had sent you to get the others..." The rest was unspoken.

Bhasin ran. He knew which side of the bread was buttered and which side was about to get seriously toasted. He ran to the first cabin which was unlocked. It was the BCA's pal and he imparted news of the impending disaster.

The pal was a bit woozy having woken up from a deep sleep but managed to make out the clear and present danger in Bhasin's trembling voice. He put on his shirt, pants and coat over his pajamas, slipped on his shoes without socks. The tie was already knotted and hanging on the mosquito net rod. BCA's pal and potential victim of Capt Jeevan, GC Sodhi bolted without even bolting the door of his cabin.

He was just in time...Jeevan roared around the corner on his bike and rode up the ramp into the quadrangle on his thundering 500cc iron steed. Sodhi vanished around the corner, simultaneously sprinting, zipping up his pant...while mentally rehearsing his alibi, "Sir my stomach was upset so I left the fall in to go to the latrine!"

GC Bhasin was unfortunately not fast enough, and ended up transfixed in the headlights like a rabbit. He looked very much like that unfortunate animal, except he was very inedible.

"What the hell are you doing here when there's a fall-in?!"asked Jeevan

"Sir...I was sent by the BCA to call all those GCs in the cabins..." said Bhasin stammered nervously.

"All right, fuck off back to the fall-in" replied Jeevan as he turned his attention to his captives.

It was a motley group of GCs, some shivering with fever, some limping because of sprains and other sports or training-related injuries. As in any place with harsh living conditions and the constant fear of retribution, there were a few guilty of 'shamming' as well. The genuine cases looked sick but not defeated, because they had immunity. Not immunity from being hauled out for questioning of course, but immunity from being physically and mentally abused in fall-ins. The shammers[2] had no such luck.

There were two unfortunates who had attempted to hoodwink the system. Capt. Jeevan took the system very seriously. The two GCs who had pretended to be ill to avoid the fall-in fully appreciated the gravity of their crime. It was stupid to presume that Jeevan would not read their medical excuse chits carefully, chits that had expired two days ago. The wrongdoing duo fully appreciated their predicament as they stood shivering in the minus temperature, clad only in their 'undies' (underwear).

Jeevan dismissed the genuine cases who had chits from the doctors attesting to the patients being SIQ (sick in quarters) or ex-PPG (excused PT, parade, games).The two shivering wretches were ordered to roll in front of Jeevan's motorcycle, like some barbarian chief's outriders.

It was thus that Lord Jeevan's cavalcade rolled back to the battalion. The sight of the two semi-naked rolling escorts struck fresh terror into the hearts and minds of the assembled GCs in various pain-inducing poses. If an officer did not mind having cadets with bruised backs to show, it meant the worst. It meant the powers that be had sanctioned the punishments and there could be no redemption or rescue.

Jeevan set the ball rolling with a relatively small moral lecture, "You fuckers are trying to fool the system and sham it out...these are totally unofficerlike qualities and unacceptable!" He paused for effect, "There are some of you who have not understood the ethos of the Army and the Academy...but after tonight all of you will! Those who still don't...are strongly advised to put in their papers and leave the Indian Military Academy!"

He then turned to the BCA and BUO, "All right BUO, I want to see the II termers sweating. BCA, take care of the I termers-they seem a bit weak in PT! III termers get in three lines..."

GC Anant walked up to the BCA, who looked a bit confused and whispered something in his ears. The BCA then walked up to Capt. Jeevan and said something. Jeevan nodded, the BCA nodded to Anant who turned to walk away. At that moment the entire battalion would have given up something really dear to know the magic words that had liberated Anant from the open air torture arena.

But Anant was not off the hook. Jeevan shouted, "Fucker! You bloody well stay and watch your coursemates suffer because of your doubts!"

So it was official, their tribulations were Anant's fault. Jay and his compatriots did not have much of a chance to speculate because they heard the BCA say, "Okaaaay gentlemen...(The pause was important) we are going to start with two hundred pushups...lancee give the count...and make sure nobody shams! Any fucker I find shamming joins Capt Jeevan's party... and that includes you, friend lancee."

Lance Corporal Pandey alias 'Lancee' jumped to it and ensured that the BCA's bidding was carried out to the letter. He would later invite his platoon I termers to his room and share a cigarette and maybe some grass. But for now it was his ass on the line. Orders were orders, thus neatly echoing what the Germans at the Nuremburg trials had said.

The I termers' collective minds boggled at the thought of doing pushups in numbers that ran into three figures. But the alternative was being delivered up to Capt. Jeevan for his advanced treatment. And the BCA had wisely channelled their rage against Anant and they thought that they were better off with the lancee.

Indeed the I termers watched as Jeevan made the third termers get into three lines. Then the GCs in the rear file were ordered to climb on the shoulders of the ones in the middle row. The cadets in the front row got into push-up position with their feet held by the middle row guys. So in a brilliant masterstroke, three cadets were engaged at a time, all in plain sight with no chance of 'shamming' it out. The GC in the middle row had it the worst since he had to go up and down carrying one on the shoulder while holding the third one's feet. But Jeevan was considerate enough to make them change position after every count of twenty.

The long night had begun...and the I termers put sincere efforts into their exertions. The first lessons of military indoctrination and powers of psychological motivation were implemented. The moral of the story would be etched into the minds of the I termers, 'Just follow orders without complaint and avoid quantum increase in discomfort.'

There were two watchers in the shadows following the events keenly. One was Bhasin, who had stayed long enough to ascertain that nobody had missed him. The only people who should have missed him were his coursemates and his master, the BCA, both of whom were painfully and gainfully employed. Both the painfully and gainfully employed parties were bound by the common aim of getting through the night.

The other watcher was Capt. Shah, there under the express orders of the CO. He did not want Jeevan to go overboard. It had happened before and the section hospital doctor had reported the high rate of injuries directly to the commandant. The matter had not been taken very seriously by the commandant, but a repeat would not help the CO's reputation at all.

Capt. Shah remembered the famous case when a third term cadet had been killed. It was a tradition where third termers could be hazed on the final day before they passed out. There had been this one cadet who had made life miserable for his juniors and they had taken things too far. It had started relatively innocuously with the cadet being forced to get ten cups of very hot tea at a time. It then progressed to beatings with sandbags, with someone using a sandbag filled with rocks, resulting in a ruptured kidney. The cadet happened to be a three star general's nephew. There had been enquiries ordered, tears...ruined lives and careers. The final result had been a strictly enforced ban on the final day '*ragda*'[3], a classic case of 'too much, too late'.

Capt. Shah stayed long enough to make sure that Capt. Jeevan stuck to the approved punishments without getting creative or into the NDA mode. The NDA was for teenage boys while those in the IMA were, for the most part, full grown men. Punishments have been a way of life in the Academy since its inception. Too much degradation and indignity, and the scope for a spontaneous resurgence of aggressive defiance also known as mutiny, increased.

Once that threshold was crossed, the word would spread rapidly and no I termer would take any orders from any senior. Civilization and grooming as the Academy knew it would be over... as would be a lot of careers.

The next day Capt. Shah went to the CO, Col. Sehgal, to give him his report. Col. Sehgal was the AD (air defence) artillery and had seen no field postings, but was determined this would not stand in the way of his promotion prospects.

The CO was apprehensive, "I hope they're all ok?!"

"Yes sir! I just checked with the RMO...the two cadets who did the bareback rolling have a bit of a cold and some bruises that's all..." said Shah reassuringly.

Just then Capt. Jeevan walked in with GC Anant. Jeevan looked smug and Anant relieved.

"Sir this shago[4] wants to leave the Army; I have brought him for your interview!"

Col Sehgal was livid, "Why? Not man enough, huh?"

Anant stood with his head down, till Jeevan tapped him on his shoulder, "Speak up, you silly bugger!"

Anant looked up with tears rolling down his fat face. "Sir I joined the IMA to become an officer but ever since I came here, I have been treated like a peon. I didn't even get to see my office. I want to go home and live in dignity...we are not so poor that we need to work in a place like this without respect..."

Sehgal looked at Shah who shrugged his shoulders. All three officers were foxed by the office part and a bit embarrassed by the sight of a grown man crying.

Col Sehgal muttered, "Respect... has to be earned, not mugged up and won once and forever like the fucking civil services... If you feel you are not fit for the Army, you should bloody well leave. Jeevan do the needful!"

Jeevan muttered "Bloody sissy...come with me and sign the paperwork for release." As they left, he threw in an extra remark, "We don't need weak will-powered asses like you!"

And thus GC Anant won his freedom from the pressures of training to become an officer and a gentleman. The gentle part of gentleman he never got to see or imagine neither did he figure out what it was like to be a strong will powered ass.

"I can't imagine what the fucks he was talking about? Office? What fucking office? These buggers come here with some very weird bloody expectations..." said Col Sehgal as he shook his head.

Capt. Shah could not disagree. He had learnt his lessons well in his time as a GC.

The II termers were still licking their wounds, having been hit the hardest. They believed the popular rumour that the I termers had precipitated the 'fuck fall-in' and now they waited for a chance to avenge the sweat spilled. The lesson they had learnt was nothing new but just a revision, 'don't fart against thunder'. The third termers had also been reminded, though a bit more forcefully than they would have preferred, 'Never catch the bull by the lull.'

These lessons were valid not only in the Academy but in any work place with a hierarchical setup.

NOTES

1. Civil Disobedience was one of the movements led by Mahatma Gandhi during the freedom struggle in the 1930s.
2. Shammer—slang for one who shams...or avoids work/responsibility or in the case of the Academy, hardship.
3. Ragda literally means scrubbing or scouring but it is actually slang for 'tough time'
4. Shago—a common slang used in the army denoting weakling or tired out, short for 'shagged out'.

Chapter 11

SOCIAL ANIMALS

It seemed to Jay, and indeed to many junior creatures in armies the world over that the rules were kind of unfair. There were two important rules for having fun, time off, breaks or recreation. The first being that juniors were created to allow seniors to have time-off or fun. This was an important aspect and should never be confused in the reverse. Those who did and were impatient for fun beyond their destiny or time paid the price with failed careers and a distinct lack of mental peace.

The second rule was also simple, if you wanted to have fun in the Army or in any hierarchical group; you had to wait till you became senior. This was a bit confusing to some of those from a thoroughly civilian background, as it was clear that there would always be someone senior to you in the Army.

This point was cleared up for the I termers by their avuncular Lancee Pandey. They were sitting on the ground, around his bed after another punishment parade thinly disguised as PT. Lancee Pandey was offering some sanctuary and succour in the form of nicotine in his room. The atmosphere was cosy and there was actually some bonhomie, which was somewhat warming up the sweat-soaked I termers. It was also heart warming since they were supposedly on punishment time though cooling their heels or warming their ass if you like in Lancee Pandey's cabin. Lancee Pandey's explanation of senior-junior relations, between puffs of cigarette smoke, was simple and to the point.

"Till such time as you have a junior of your own... you remain a bloody junior!"

There was much mental mastication of the said information. Finally a GC of deep intellectual capacity and infinite capability to screw up simple situations popped the question. The question was well-meaning and arose out of genuine ignorance. But it pricked the bubble of good feeling created by nicotine and a benign senior.

"But sir...that means even if you are senior but don't have any juniors...for example, does that mean you're junior when we are not around?" This was followed by confused and apprehensive silence, as the audience struggled to decode what had been said. They also struggled to control their horror at the faux pas and its possible repercussions. They realised the statement was possibly offensive even before they fully understood it. They were beginning to comprehend Army etiquette and the perils of saying something a senior might not like.

Lancee Pandey was not quite sure what had been asked. But it sounded like dissidence. He looked up irately to see who was spoiling the congenial '*guru shishya*' [1] atmosphere. One must not ask difficult questions in the Army, especially during informal sessions such as this. For a few moments Pandey had also been able to forget his own wretched existence as overseer and had actually begun to warm up to the slaves. Now he looked around suspiciously, trying to ascertain if the I termers weren't trying to get extra 'pally' or generally act 'funny'.

The question had been very close to the truth, but the truth can be bad for morale. Truth and morale don't go together either in the reporting or in the maintaining of it. After some tense moments for the I termers, Pandey realized it was a genuine question.

"Listen, you bugger...it's simple... whoever is at the bottom of the heap at that particular point is the bloody junior," and a big cloud of smoke issued forth from Lancee Pandey's lips, almost in the form of an exclamation mark.

The gesture was clear to most, the hidden meaning was clearer to all, 'Enough bloody questions!'

Jay felt someone tugging at his sleeve. It was Shinde trying to pull him out of the Lancee's room. They managed to sidle out just in time as they heard Lancee Pandey say, "Ok fuckers get on your hands down and give me thirty...that'll get you warmed up for the social and maybe clear doubts about seniority or why your junior ass is mine!"

That evening there was a company social. It was supposed to be a secret but the whole place was abuzz with a kind of repressed excitement. Everyone, including the drill instructors and officers knew what was going to take place. But to a casual observer there would have been no signs that an impending orgy was to commence in the next few hours.

It was an old custom to have these platoon or battalion socials, usually to commemorate some important landmark like finishing off a camp or test. The name 'social' was innocuous as compared to the actual chaos that ensued after a few hours. There would be no semblance of polite society or civilisation. Thus, most instructors, even the duty officers and NCOs stayed away; it was better that way for all concerned. Today's bash was for a reason widely and often quoted in the Army circles, 'Just out of sheer bloody happiness!'

The orderlies had already managed some bales of hay from the stables and made stacks in the centre of the quadrangle along with whatever wood they could scrape up. The more intrepid of the GCs had managed some liquor. There is nothing more fun than watching some virgin drinker run amok. Liquor always lubricated the social part of the 'social' anywhere in the world. The equitation officer would have been incensed if he heard about his animal's food being pilfered. It was understandable, after all that man loved the horses more than the cadets. The DSs understood and generally looked the other way.

The guys who had managed the alcohol were mostly GCs who belonged to

Dehradun, or were from the ranks or ex-ACC. The last by virtue of being ex-soldiers, still had affiliation with the NCOs from their regiments who were now instructors. Everyone was willing to oblige for a good cause. By and large, since these were GCs who would soon become officers, it was a given that they would be responsible citizens and drinkers.

It did not always happen that way. Certain excesses under the influence of alcohol, in the past, had pressured a previous commandant to ban these unauthorized and often unmilitary gatherings. These 'excesses' were mostly fist fights, wrestling and maybe the occasional destruction of government property. It may have been a bit unmilitary but nothing really unmanly. But there had been this one particular case when all limits had been crossed and the cadet appointments had imposed a self-ban with the approval of almost all the senior GCs.

Socials had been banned after some platoon had burnt all their furniture and then proceeded to pee in the drill square. The furniture eradication program could have been forgiven, but not the sacrilege of soiling the blessed drill square. The square was hallowed. Generations of soldiers and martyrs had shed their sweat, sometimes blood and other fluids on that square. Nonetheless, these unauthorized military orgies restarted, after the constipated commandant was posted out and the third termers had passed out. The present lot of third termers had decided to reinstate the socials. And that was the reason for this social. Whatever the reason, it was welcome.

For their social, they were permitted to wear civilian clothes. Jay put on his jeans. They felt weird after two months. They also did not feel as if they belonged to him. In two months he had lost many unwanted kilos...kilos that only a fond and indulgent mother would lament. Jay sadly looked at his reflection in the mirror.

The reflection that looked back him was just as sorry looking...a sunburnt face with close-cropped hair on top, machined on the sides and back. The civvies were hanging on him making him look like a scarecrow. Col. Shankar would have been proud and would have said something like "Here at last is a lean, mean fighting machine." Mrs Shankar might have disagreed, but mothers are irrational in matters regarding a progeny's adipose ratio.

Jay looked around for his belt desperately. He needed it badly; it was the only thing that would hold up his pants and his dignity that evening. After having dropped two sizes, the jeans were more influenced by gravity than by his hips. He could not find it and realized one of GCs might have '*maroed*'[2] it. There was nothing to replace it as his spare belt had been 'borrowed' by his neighbour and 'in loco parentis', GC Corporal Rathore. Jay was beginning to realize, borrowed equalled *maroed* in the Army, if a senior was involved. But time was a-wasting and so he just took the 'line bedding'[3] and tied it around his waist. He was just in time as the first roar of the fire was cheered with much hooting and whistling.

He went outside and bumped into the BCA, the resident superstar. Jay was appalled, but the BCA giggled drunkenly and said, "Fucker you're too ugly or I would have asked you to meet me in 'undie order'[4]!" And with that he fired off a mini flare even as Jay winced.

This military disorder made everyone seem more human and Jay suddenly loved the Army. Everything and everyone looked rosy, warm and cuddly. A few shots of rum later his memory became as hazy as the situation rosy and despicable seniors cuddly. Everyone was friendly and approachable.

Suddenly there was no hierarchy, only soldiers and comrades who would give their lives for each other in battle. There was the 'evil' JUO Akshay Kumar doing an extremely risqué cabaret dance on a table dragged out by his sworn enemies, the II termers. Many blood oaths of loyalty and assurances of providing aid and rescue when surrounded by hordes of enemies were sworn that night.

There was food and lots of drink and maybe no tomorrow. Tomorrow was a Sunday... who gave a damn about anything! The punishment parade types came in later, after the night report of their punishment parade. They were welcomed like heroes back from a special mission. The punishment types had a rebel air about them.

The party had everything a good orgy should have, except for the absence of women which everyone missed but no one really minded. It was universally acknowledged that they would have mucked up the atmosphere.

Then the eponymous song 'You're in the army now' by the band, Status Quo, was played. All the cadets stood up and cheered. The song was played again and again and then again. Everyone sang along and screamed the lines '....Stand up and fight...you're in the Army now!' Jay and some of the others, suitably motivated, and inebriated, had tears in their eyes. They swayed out of sync with the tune and screamed again and again "You're in the army now...oh hoo ohhh you're in the Army now."

At some point inspite of the BCA's half-hearted and the BUO's frantic protests, more furniture was dragged out and burnt. But to be fair that was only when the wood and hay had run out. GCs were lying asleep or inebriated, mostly both, in various postures, in assorted places. The combination of youth, sleep deprivation, a clean conscience and a little alcohol makes for very deep sleep indeed.

The BCA and BUO however were somewhat worried men, but not too much, thanks to the booze. They had tacitly taken responsibility for any mess that happened in the social. But now it was late in the night and everyone had turned back into pumpkins, rats or whatever they had been before, and now there was no one to clean up. The BUO told the BCA "Ok partner, pass the orders for cleaning this mess..." and passed out!

'Never mind!' the BCA thought to himself, 'We'll get the I termers to clean it up tomorrow...and the orderlies to mop it up on Monday.' It was his last coherent thought of the evening. He also said it loudly. It was heard, recorded and noted for transmission the next day by the infamous understudy GC Bhasin, who was unofficially unwelcome at the social. Being unwelcome he had been lurking in the background.

Bhasin was feeling a bit guilty about having missed all his coursemates' tribulations and hence watched the revelries from the background. To be fair he also watched them undergo punishments from the background, from a safe distance. However he heard his master's words and faithfully recorded them. He also helped the staggering BCA to bed. He was loyal to his master, if not to his mates. He would try the social animal bit later when survival and comfort was not an issue.

NOTES

1. Guru Shishya—teacher student relationship
2. Marrowed is slang for steal or swipe, from the Hindi word 'maro' which in the strict sense means—to hit or kill.
3. Line Bedding A part of regulation pack contents, basically a length of hemp rope.
4. 'Undie order'—dressed only in underwear or briefs

Chapter 12

SUNDAY AND UNWASHED MASSES

Sunday dawned bright and clear, unlike the hung-over GCs. A I termer came and made an announcement at 0630 hours. The announcement was made in the particular sing-song way that was a tradition in Jay's platoon. This quaint tradition had been introduced by the ex-NDAs to IMA. The announcements may have looked adorable in NDA when delivered by a cute beardless high school graduate. It did not carry the same charm, when a husky and hairy college graduate in his twenties, delivered the message. Someone had deliberately misled the poor announcer about the tenor. It succeeded in making the formal announcement sound ridiculous.

"All I termers fall in at 0700 hours in PT rig (*pause*) to clean up the mess (*musical pause*) under arrangements of platoon sergeant!"

This poor wretch of an announcer was greeted by shoes and assorted refuse from almost all cabins and all courses, senior or junior. Sadly, he had to go through this at all platoons, because the content was serious, even if the method of delivery was not. He very nearly got a permanent treble, because of the verbal assaults on his sexuality and physical assaults on his person.

The I termers hated the announcement and hence the messenger of bad tidings. After all they would be the ones doing the actual clearing up. The II termers could not leave till the mess had been cleared up. The III term GCs and the powers that be were not morally free to go on an 'out pass'.

The III termers were apprehensive, with the exception of a few 'don't give a damn' type souls. If they left without assisting in some way, some duty officer might visit and finding no 'team spirit' amongst the III termers, get upset.

That might mean a very big upset, to an otherwise outstanding weekend thus far.

The I termers had only one aim on holidays and that was to escape or at the very least, to mitigate any attempt to extract services from them. To that end, most of them had gotten ready very early in the morning, locked their cabins and then re-entered from the back door. They then spent the next few hours catching up on their sleep.

The wise even missed breakfast in order to catch up on their sleep. This was most important if one wanted to avoid being press-ganged in the dining hall. The wiser slept just next to the door to avoid overzealous seniors. These seniors were prone to check if the I termer was actually in the room. This was done through the simple expedient of peeking through the ventilator. After all, the seniors had done the same thing in their junior terms.

The ventilator was right above the front door and hence anyone lying next to the door could not be detected. A determined search, with mirrors for instance, would soon give the game away but that was only on rare occasions. And it was only done by the rare 'psycho' senior. There were several of those so it was a matter of some concern to the I termers. But luckily there were more I termers than psycho seniors so it was just a matter of chance.

But there was no choice for at least some of the I termers. Most of them could not make any plans for some R&R (rest and recreation) without getting entangled in menial tasks like getting tea or hot water. The plans were simple and not very inventive, mostly involving going on outpass, ogling girls, sleeping and eating good food, and not necessarily in that order. The I termers, especially those who hadn't passed their drill tests, had very little choice. It was usually hide, sleep and eat at the café or shopping centre, hide and sleep again, usually in that order. Jay decided to take the rare third option and watch a movie in the auditorium.

But the '*Ghadr*[1] party' had other plans, other, but essentially simple plans. They were all from the same school in Punjab and were very tight-knit. The knit stood the test, especially in matters concerning hoodwinking authority and having a raucously good time. They had been nicknamed the '*Ghadr* party' after a freedom-fighter type of the same name in Punjab half a century earlier. They had been known for their disruptiveness during the Indian freedom struggle

The present *Ghadr* party had no qualms in having fun at the expense of others' freedom. They usually stayed in their cabins, drank and ate butter chicken, in short, recreated the ambience of Punjab. The *Ghadr* party was a bit aggrieved at being denied their fundamental right to speak in Punjabi and sit in a group in public. In the Academy, Sikhs, the most visible sign of 'Punjabiness', were not permitted to sit together in numbers more than two. This rule was to prevent communalism, regionalism and groupism; all three 'isms' being core values of the Ghadr party.

The rule was usually circumvented by making a 'cut surd' (a Sikh, with trimmed hair and beard and thus no turban to give him away) sit in between. Any other 'lesser' Punjabi speaker would also do in a pinch. They just loved to sit together in a group. Probably the same as when they had been when fighting the Mughals, but Academy rules were firm. It did not matter that no one wanted to sit amongst them. It was pointless, since the conversation was almost exclusively in Punjabi.

This group which consisted of all 'terms', senior and junior, was united in their common belief of the superiority of Punjabi culture. It must be conceded that they managed to have a lot more fun than other groups. At least definitely more than any other regional group like the Jat *bhais*, '*Tants*' (Maharashtrians) or Bihari *bhaiyyas*.

They looked forward to Sundays when they could sit and chatter, drink, eat and get drunk all in Punjabi. They also looked eagerly for chattel who would be able to serve them while they enjoyed themselves.

The requirement was for someone who could be sent to get the required soda or snacks from the shopping centre and the butter chicken from the café. The orderlies would also do it but they would expect tips and very few turned up on Sundays. Anyway, if there was one thing the *Ghadr* party was good at, it was living off the land. Why pay a tip when you could get it done for free. The I termers came free, but finding one at eight in the morning was like getting a cab during rush hour. They had I termers in their club but it would have been incestuous making them run errands. No...they had to find non-Punjabi races, preferably servile ones.

They sent out a pet I termer and active *Ghadr* party activist, GC Chiranjeev Bansal on a mission. His job was to infiltrate and lure a fellow I termer to the *Ghadr* party HQ, where the victim could be further tasked. The logical solution would have been to utilize the services of GC Bansal, a I termer himself.

But GC Bansal, being from a rich family in Jalandhar with deep pockets was one of the main sponsors of the *Ghadr* party revelries. He was thus approved pimp, procurer and Quisling[2], roles he carried out with remarkable relish and gratitude. Not to forget a zeal that transcended mere survival instinct. He was from a business family. He was paid in kindness—he was protected from exactly the same life of servitude he was about to entice another I termer into.

Most of Bansal's coursemates knew of this 'human Venus trap' and avoided him like the plague. But there were still some who would get carried away by the milk of human kindness and course spirit. Every weekend most of the I termers managed to evade Bansal. But as someone in Hollywood once said, 'There is a sucker born every second'. He might have been talking about the IMA; there was one born that weekend in Bhagat battalion too.

GC Rohit Kapoor was loitering in front of his room looking for his keys. The keys were in one of his many pockets. This he would realize at some point. He had just had a busy morning and was not thinking clearly, which was the norm, anyway. He had realized in a vague kind of way that he had made a mistake hanging around the bathroom. Wishing the BCA a cheery "Good morning" had not been a very good idea. The BCA had had a hangover and had replied to Kapoor's good cheer with a snarling, "What the fuck are you doing dawdling around the fucking bathroom? Report to the CSM and help clean up last night's mess!" With that the BCA poured a mug of cold water over his head.

GC Rohit Kapoor tried to help by saying, "Sir, I have a better idea, try hot water..."

Thirty pushups later...on the wet bathroom floor, GC Rohit Kapoor learnt the value of holding onto unwanted counsel...especially in bathrooms. Two back-breaking hours later he also vowed never to hang around seniors with hangovers. The CSM had also been doubly cranky because of a killer hangover and the mysterious disappearance of almost all I termers. Nevertheless he was a decent soul and did not keep Kapoor toiling for more than two hours.

Kapoor was looking forward to a well-deserved nap, a leisurely hot bath and some good food at the café. He had just got some money from his dad; he could afford to treat himself. What he could not afford was dawdling around in plain sight. He was still fumbling with the lock, when Bansal who had been lurking in the shadows, spotted him.

Bansal had craftily positioned himself near the bathroom. His reasoning was simple and similar to that of any predator in ambush near a watering hole.

Just as the odds of prey approaching a water hole increase, it also followed logic that the call of nature would inevitably draw some I termer out of hiding to the bathroom. And thus it happened that Kapoor fell into Bansal's slimy paws.

"Hey Kapoor what's up? Free *ho gaya* finally?" asked Bansal with a fake sympathetic smile as he moved in for the kill.

"Yeah, silly bastards used us like slave labour moving all that stuff from last night! Where were you old boy?!" replied Kapoor.

"Man! I was slaving for that Jalandhar crowd (pregnant pause)...by the way...they told me to call you?"

The last bit penetrated Kapoor's exhausted mind. He whirled around and looked for Bansal who had retreated to a safe distance out of punching range. Kapoor wrestled with his conscience. Bansal had realized there might be some reluctance. He rushed off and informed his clients about the presence of a candidate, who might need some persuasion.

Bansal was eager to recruit Kapoor, as the alternative was unthinkable. He would have to take on the medieval role of serving wench and waiter to the ravening group of half-drunk Punjabi seniors. GC Amarjot Sandhu came to Bansal's rescue. He came out of the party cabin briefly to yell for Kapoor.

Kapoor's mind was made up for him as he trudged wearily to the HQ of the Ghadr party. He was promptly given a list of chores and told to hurry up. They included getting two jugs of water first and then going to the shopping centre and getting snacks and some Coke ostensibly to mix with the rum.

Kapoor stood near the water cooler and mentally ripped up his plans for a good Sunday. His good-natured mind was slowly coming to the realization that he might have been 'framed' by his comrade Bansal. He heard an urgent hiss from a nearby cabin. It was Ghai looking at him through the ventilator, and gesticulating urgently.

The cabin happened to belong to Jay but was currently acting as a shelter for abused I termers. The location of this sanctuary was eagerly sought out by almost all II and III termers and some I termer turncoats like Bansal or toadies like Bhasin. Luckily none of the interested parties were in the vicinity or they would have found Kapoor's reactions very revealing. That is revealing a nest of I termers who could be gainfully employed. But Kapoor was now wiser and got the message. He went around the back to access the cabin's back door discreetly.

There he was asked by Ghai, "What the fuck do you think you are doing, you *chutiya*?!We're waiting for you so we can get to the café!"

Kapoor winced and poured out his tale of suffering and exploitation. But there was more to come as the full details of Bansal's perfidy became clear. Kapoor was enraged at this betrayal by a comrade. He still believed strongly in course spirit and loyalty. The others hiding in Jay's cabin, all six of them, could not understand how anyone could be so naïve. Kapoor was naïve but could also be quite dangerous when aroused. Just then Bansal came out of his cabin to add further impetus to Kapoor's righteous ire.

"*Oye* Kapoor...where's the water, the seniors are waiting!" Bansal had been emboldened by alcohol and thus had blown his cover. He had thus far pretended to be toiling for, rather than working with or for, the *Ghadr* party. The distinction was significant, and life threatening.

Kapoor blew his top and had to be physically restrained from charging out of the cabin's front door and blowing the safe haven of the I termers. He calmed down and went out the way he had come in. He did not forget the water jugs.

He smiled at Bansal and said, "Have a good weekend, motherfucker!" He then casually and coolly poured the water over Bansal's head and walked off. Bansal stared after him in shock.

And that was when Bansal's pals came out. One of them, a dishevelled Sikh named Hundal, was slightly inebriated by party norms. He shakily started towards Kapoor issuing threatening statements in an indecipherable language possibly Punjabi.

Kapoor was not in any mood to be pushed around anymore. He threw the empty water jugs at Hundal. It also sounded like he said something that sounded very suspiciously like "fuck you" to the I termers watching discreetly from Jay's cabin. The members of the *Ghadr* party were incensed. There was no doubt that if he had been in Punjab he would have been severely mauled by the 'Shers e Punjab'. Some forceful statements to that effect were issued by said *shers*.

The CSM came there and stood looking at the tableau without saying anything. Nothing needed to be said, this was the IMA, and ultimately wisdom prevailed. It prevailed under the watchful gaze of possible retribution aka the CSM. The more sober members of the *Ghadr* party dragged Bansal and Hundal back to their cabin. A few foul and choice Punjabi abuses were half-heartedly hurled in Kapoor's direction. When faced with the first sparks of mutiny it is better to stamp it out or as it was in this case, not let others see the embers.

The others had witnessed the drama from the safety of the cabin's ventilator. The time had come to escape before Bansal remembered from which direction Kapoor had made his appearance. It would not take too long even for a bunch of half-drunk II and III termers to put two and two together. Jay and the gang snuck out and headed for the hills, literally. Then Jay decided to go back for Kapoor. He was just in time—Kapoor was thinking of going back for Bansal. Jay managed to persuade him to follow without another thought of bathing or other such allied traditional Sunday activities.

Ten minutes later they were in the Tons River training area, sleeping happily in the sun. They felt the sun on their face, dirt in their hair and grime on their semi-washed bodies...but their souls were free and happy...

NOTES

1. Ghadr party is a reference to a group that was part of the Indian freedom struggle. The word 'Ghadr' also loosely means chaos, which is what it refers to in this case.
2. Quisling was the infamous Norwegian collaborator during the World War II with the Nazis during their occupation of that country. His name is synonymous with betrayal and collaboration with the enemy.

Chapter 13

Manic Mondays

Blessed Sunday was over, the one single day when they could leave the Army behind...and get back to humanity. Jay curled up in his thin IMA issue grey, filthy blanket and tried to postpone the dreaded hour of rising. A quilt would have helped him feel warmer and cosier, but it was not regulation and anyway difficult to dispose of during 'cabin cupboard' inspections. He heard the bugles, sirens and other sounds the army made to rouse its denizens. There was muted rustling, as the cadets moved about in the semi-darkness trying to get ready.

Jay did not give a damn or he tried not to give one. He was sleep deprived, still slightly culture shocked even after a couple of months and very, *very* tired. The few hours on Sunday had not made up for the exhaustion of the past week. Right now, sleep seemed the best cure. The cure for all the acute symptoms of artificial poverty and the lack of mother's love he was feeling.

He heard a loud knocking on the door and tried to block it out by burying his head under the pillow. An image of himself sweating under triple pack *bajri* woke him up rather conclusively. But he did not open the door and waited hopefully till whoever it was had gone away. Jay was not in the right frame of mind to run errands for some senior this early in the day, especially today.

Jay was also acutely aware of the fact that this was one of the 'double jeopardy days'. It referred to the days when the first two periods before breakfast were drill and/or PT. It was when a period of drill was followed by PT, that a frantic rush took place to be 'in the right place in the right time in the right rig'. This was supposedly the magic mantra for a happy and prosperous military career. The I termers were still struggling to cope with the frenzy that was essential to make it on time...most did not.

The drill and PT instructors understood the plight of the freshmen, but could not show leniency or pity. The army is NOT a charitable organization except for generals and their progeny...at least that was the popular legend. Actually, barring a few sadists, the instructors were generally less harsh with the I termers. For example for being late, a I termer could expect to get off with a few minutes of punishment, while a II or III termer would probably be methodically reduced to a lump of exhausted meat. He would probably also get a few restrictions or ED (extra drill) parades to boot...pun unintended.

Jay was aware of all this, but not consoled by any of the aforementioned general knowledge of the instructor's lenience. He would much rather be on time, and so he rushed to the bathroom to get ready. Taking a crap was not important; shaving was. That is if you did not want to get the crap drilled out of you. Going unshaven to the drill square was asking for misery. I termers had enough misery on their plate.

The fates determined otherwise. He had almost reached the safety of the bathroom when he heard someone yell out. In the early morning darkness he could vaguely make out someone in a dressing gown crossing the quadrangle. Jay did not need some senior wasting his time with some silly errand at this time. Most seniors did not give last minute errands to I termers especially when they had drill in the I period. But Jay suspected this was Iyer, a particularly irritating and perpetually whiny II termer. Confident of not being recognized in the dark, Jay quickly dashed to the bathroom where he would be lost in the crowd of cadets of all terms zipping through their toilet routines.

Jay made a mistake-he looked over his shoulder. He could make out Iyer's head and shoulders sticking out of the centre of the quadrangle. It then hit him. There had been a block in the latrine, and they had dug a pit in the centre of the quadrangle to clear the blockage in the sewage line. The blockage had been cleared but the workers had left a three-foot deep pit full of cold shit. It was into this pit that Iyer had taken the plunge. Iyer was in deep shit but that was of some cold comfort to Jay. That was when Jay made his second mistake. Feeling pity-the last remnants of his pre cadet stock, he rushed to the pit and asked, "Sir, any help?"

Iyer replied with the kind of disjointed invective as could be expected from someone waist high in cold excreta.

"Yes! Get me the hell out of here!"

With extreme reluctance, Jay got hold of Iyer's shitty arm (pun intended) and pulled him out. This took some precious minutes and Jay was aware he was running out of seconds. Iyer was finally extricated and Jay tried to slip away but Iyer called him back.

"I've dropped my mug in there get it for me, I'm going to the bathroom, hurry up and bring it there..."

Jay marvelled at the dastardliness of this sonavabitch. If Iyer could only have seen the disgust in Jay's expression he would have thought twice before turning his shitty back. The disgust was not at the thought of fishing out the mug from the shit. It was at that small-minded ingrate Iyer who would want to use that mug again.

Jay wrestled with the tantalizing option of tossing Iyer back in the pit but he did not have time. He wanted to scream 'Fuck you' but instead mumbled something and waited for Iyer to leave. He wanted to run and hide but opted to run to the bathroom and finish his 'shit and shave'... 'shower and shampoo' were civilian luxuries. Right now it was important to get away before Iyer spotted him again and asked for that precious mug.

The worst part about going off schedule at the Academy was the loss of a squad. As mentioned earlier only cadet appointments were allowed to ride alone. Without a squad, a cadet could not ride his cycle and would thus be forced to run with his bike, and usually end up late. That would then trigger a chain reaction of late arrival and punishment, some summary and some official. Summary justice in the Army meant something on-the-spot, quick and painful, physically exerting and not always regulation.

Official punishment on the other hand was recordable in your dossier and while not so painful, made up in terms by time wasted. The accumulation of official punishment usually took cadets their whole Academy stay to work off. It could also lead to relegation. The effects of summary or on-the-spot punishments could be recovered from in 24 to 48 hours.

Jay knew that and felt his heart sink when he finally rushed out and saw the last of the squads departing. He rushed to get his bike and catch up with them but was stopped dead in his tracks by the sight of an appointment lounging around near the battalion *chowk*[1] . There was no way he could avoid him. Jay decided to abandon the bike and started running.

The drill square was milling with cadets attempting to get into some orderly formation. The adjutant was on his horse, looking eagle-eyed for an offender. The assistant adjutant who was eager to be more feared than his boss was also around harrying squads. He was disliked by cadet and drill instructor alike.

The drill period had not begun, but the cadets were already sweating as they were 'warmed up' by angrily barking drill instructors. Jay somehow managed to avoid the adjutant and his lackeys and slipped into his company squad. He was welcomed by his drill senior JCO who was also eagle-eyed but more compassionate than most of the officers. Since there were still four minutes for the drill parade to start, Jay was not actually late, but as far as the drill square etiquette was concerned Jay was one minute late. One minute was...very late. He was let off lightly with the order "Twenty 'flat foot'[2] carry on GC!!" The drill instructors loved to give commands in English. It made them feel like worthy representatives of the officers' training team.

The rest of the drill period went off relatively uneventfully. Jay was part of the majority that had failed the I termers drill test and were thus deserving of some intensive coaching. They even had a large drill test number stencilled on their breast pocket. It just added to the feeling you could never escape this prison. The coaching was by the company drill instructors who considered the cadets failing in a drill test a personal affront...no fury like an affronted drill instructor...there...you had a peeved prison guard too.

The coaching for failures seemed to focus on warming up for the first 30 minutes and then practising the drill test routine in the last 10 minutes. So in spite of the freezing early morning temperature, the drill square seemed to have a haze hanging over it caused by the steaming breath of about 400 sweating GCs.

The warming up gave them a one-minute breather, and they formed up in twos. They went through the test routine which was fairly simple. March fourteen steps, turn and march past, give a left salute, then a right one and picked up a rifle. Unfortunately that was when the adjutant cantered past. He swung his horse around and watched a few of the cadets go through their paces. They were Ghai and Kapoor, who had been brought into the world with two left feet, the grace of a pregnant dancing bear and the same martial deportment.

The drill instructor knew the outcome, a demonstration of Ghai and Kapoor's drill skills viewed by the adjutant would be very skewed and rhymed with something the victims would be. The good man hurried forward and tried to hustle the duo to the back of the line-up. But the adjutant was having none of it. He motioned the drill NCO to one side and gestured for them to start the routine. The horse snorted its approval steamily. Ghai and Kapoor started off their jerky marching that had the drill instructor wince. Even the adjutant's horse seemed embarrassed and stood with its head down. To make matters worse, the assistant adjutant sauntered up to watch the fun.

The assistant adjutant was a huge, lumbering six-footer with a deep bass growl to match his irritable demeanour. The adjutant, at this point, had had enough and being a busy man said cryptically, "Okay more practice, very poor standard!"

The squad of I termers heard this and groaned mentally in unison. Now that the drill instructor's abilities have been challenged, in front of everyone, including other drill instructors there would be hell to pay!

The I termers were right. The assistant adjutant also fondly known as the 'ass adju' added fat to the fire by adding his parting shot or the coup-de-grace, "*Aur* warming-up *chahiye*...bloody poor standard!" If there was ever something as pouring fuel to the fire that was it.

There were only five minutes left in the 40 minute period. A lot of 'warming up' could be transmitted in five minutes. The instructors and cadets knew that there were 15 minutes between the first two periods or drill and PT. Eating into those precious minutes was serious punishment. It was usually done by the meanest of instructors. And that too only under very dire circumstances like the loss of a bayonet or some other weapon accessory or part.[3]

Dire personal affront also warranted eating into the 'prep time'-the drill instructor Havildar Kishen Lal Dogra of the 10th Jammu and Kashmir Rifles regiment had been so affronted in front of his pupils. That evening he would be taunted by his fellow instructors at the NCO mess on the poor showing of his squad. And this validated by no less a personage than the academy adjutant, the demi-god of drill.

The next few minutes were the longest Jay and the bones below his knee cap, along with the muscles in his ass, would remember. They marched and sprinted around the drill store, did quick marches @ 140 steps a minute[4] and were absolutely at the end of their tether. The siren rang indicating the end of the period. The more painful part of the 'cooling down'[5] was watching the other squads move out.

Their drill instructor was an old hand, he knew that keeping the squad beyond a certain limit might mean difficult questions to be answered, but an extra two minutes would do the trick. And so it was that after two lung-searing and teeth-jarring minutes of pounding on the drill square, they were let off.

They rushed to their cycles and Jay realized he did not have his bike. He was too exhausted to run and was close to tears at the thought of more punishments for being late for the PT period. It was then he noticed an unlocked cycle with the Alamein company colour on its mudguard. It had to belong to a very senior cadet appointment; no one else would have left the cycle unlocked or be able to ride alone. The repercussions, if caught, would be enormously painful but Jay was desperate.

He had moments to act before the rest of the squad moved off. He pulled out the cycle from the stand and mounted up as Ghai rolled his eyes. He was just in time, because as Jay's squad fled, he took a backward glance and saw a cadet appointment wearing black shoulder boards looking around the cycle stand. Jay hoped his victim would not be able to make out his company colour on the cycle.

Jay and his band of unhappy men reached their cabins with just three minutes to spare. Jay rushed into his cabin and tried to rip off his ammunition[6] boots and his sweater at the same time.

He rushed to the PT ground still struggling with his white PT sweater. He made it but was made to step out by the PT NCO, a pugnacious-looking ex-services boxing champion. Jay was one of the minority who had not folded his woollen PT socks in the regulation pattern and had his shoe laces twisted. The socks were supposed to be folded around the ankles in three folds and the laces flat and untwisted. And that's a difficult thing to achieve when dressing and in a hurry to boot.

And boot was what they got from Havildar K.K. Swamy, their PT *ustad* extraordinaire. The initials KK were a mystery and the subject of much speculation. The defaulters were ordered to stand separately. Havildar K.K. Swamy set the tone for the next forty minutes with a menacing statement, "*Mein kala* cobra, and *kaatega* you GC *mar jaega...*" (I'm like a black cobra, and when I bite, you GC die)

Ghai commented sotto voce, "I thought KK stood for kala *kutta* not kobra!". Kapoor made the mistake of laughing and tried to cover it by coughing. K.K. Swamy, dark and wiry, darting around in his red stockings looked remarkably like some exotic lizard. The bark was unmistakably canine though. The others in the squad were not in a laughing mood. K.K. Swamy was also not amused. The next forty minutes would take away any reserves of mirth, or so K.K. Swamy promised the group.

The promise was delivered on thoroughly. The first order to go around the tree at the end of the PT ground was expected. The wheel barrow[7] and the crab[8] which came next were also expected. But the squad was not ready for what came after that. K.K. Swamy was after all a trained PT expert and so took his time warming them up before serving the main course.

They were divided into teams of three to carry a log and race around the blessed 'tree at the end of the ground'. The log in itself was not very heavy, but when a group comprising three different persons with different gaits and heights have to carry it, it becomes very tiring very fast.

They went through a couple of rounds of that and then moved on to the fireman lift. KK Swamy may not have been an exotic black cobra but he sure was an artist in painting the PT ground with the sweat of cadets in interesting ways. There was barely a few seconds pause before the next exercise was initiated. The flesh was very unwilling and the spirit totally missing, it was just the heart that kept them going. All thoughts left Jay's mind and the squad was like a machine running low on power. He could actually hear the seconds ticking away as he went through the motions of whatever he was ordered to do. And finally the blessed siren sang again.

K.K. Swamy said with a surprisingly friendly smile, "Last man on PT ground do fifty pushups or feel my sting!"

And Jay somehow gathered up that last effort and sprinted to his room. When he got there without being stung, he could see black spots dancing in front of his eyes from exhaustion. Maybe that was the effect of PT Ustad K.K. the *kala kobra* Swamy's venom. He dropped on his bed and was woken up 15 minutes later by a banging on the door. It was Chib. He rushed in looking worried and said, "There are some Alamein company types in the battalion area hunting for a bloody cycle".

That news galvanised Jay as if someone had doused him with a bucket of ice-cold water. The charge for stealing or 'borrowing' was pretty much the same, instant withdrawal from the Academy, in disgrace. That was after the cadet appointments had finished with him. The owner of the cycle was justifiably worried as the loss of cycle meant payment and probably some punishment for being 'irresponsible'.

Jay jumped into his uniform without bothering to change his sweat-soaked underclothes and crept out looking for the bike. It was still foolishly parked in the middle of the quadrangle. Jay had to move it out of the company area before it was spotted and the whole company was interviewed, forcefully, till someone confessed. Chib volunteered to be the lookout as Jay took the cycle and parked it in the mess cycle stand.

The mess was shared by three companies and so it would be difficult to pinpoint the culprit. They were just in time, a black tabbed III termer, wearing the Alamein company collar badge, rode by the cycle stand. At just that very moment GC Lal Dev reached the cycle stand and finding his tyre deflated, hopped on blithely onto the first unlocked bike he found, the very same stolen bike.

The black tab watched with a reddening face. He could hardly believe his eyes and moved in for the kill. Jay's faith in heaven and divine justice was revived. He went in for breakfast with a light heart and felt ready for the day and any punishments it might offer. The last image he had in his mind was that of 'Red God' cream-rolling[9] on the road. It had been kind of fun to watch...while hidden in the bushes. Jay did not know it then but the real fun would come next week when the inter company matches started.

NOTES

1. Chowk = square/ crossroads.
2. Flat foot, a warming-up routine also used as a punishment. Involves jumping high and making the knees touch the chest and then crash hard back on the ground, absolute murder on all bones below the knees, a major cause of stress fracture.
3. Loss of a weapon, ammunition, accessory or part is considered a very serious crime, and usually a career killer for somebody and varying degrees of discomfort for others, including the commanding officer of the outfit. This system has been in vogue for more than a hundred years.
4. The speed of 140 steps a minute is reserved for the rifle regiments and very difficult to maintain for the untrained and the unfit.The regular speed is 120.
5. Yes that was an important part as per regulations and just as exhausting as warming up.
6. Ammunition boots so called because the legend goes they were hard enough to be used when the ammo ran out-it had 13 hob nails and a steel shod toe and heel and felt like one was wearing a metal boot. A kick with these boots on would put a mule to shame.
7. Wheelbarrow—a PT exercise that involves one person holding the other's legs by the ankles while he walks on his hands face down-there was the other version called the spider.
8. Crab—moving forward with one's arms behind one's back, palms on the ground and feet on the ground-face up (I include this because a lady asked).
9. Cream roll involves one front roll and one back roll consecutively and continuously. It has the double effect or inducing giddiness and bruising neck, back and head and sometimes even remorse.

Chapter 14

FUN AND GAMES

Jay was screaming at the top of his lungs. So were the rest of the I and II termers of his company. Inter-company hockey matches were on and Jay was seeing a whole new aspect of sporting instinct. The academy believed firmly that victory belonged to the side with the most motivation and hard work. This readily translated into the theory that games would be won if played like demons with their arses on fire...skill was not really important. Just motivation! Talk about 'hard won'!

Defeat on the playing fields was because of only two factors, lack of motivation on the field and inadequate cheering or motivators. Freud may not have agreed with this analysis but then he had never been to the IMA. Defeat was not an option as it invariably meant the entire company being punished. The army believes very firmly in only two categories winners and losers. There was no gold, silver or bronze medal in war. The only runners up were dead or prisoner. The players of the losing side usually went on a long route march with double packs for not showing commitment. The rest of the company underwent organised 'fuck fall ins' for not supporting their team by cheering enthusiastically enough.

Jay and the rest of the Singarh Company knew the implications of defeat and were cheering lustily for their team. This was the only place on Earth one could hear cheering that had a note of sad desperation. The I termers were already hoarse from shouting the drill count during the morning drill period. But now they were exhorted to inhuman decibels levels by the III termers pacing up and down the ranks. The II termers also joined the cheering whole heartedly like their future depended on it...the immediate one certainly did.

Every now and then a less enthusiastic cadet would be yanked out and sent for a quiet round of pushups and the like. This had a salutary effect on the volume control and ensured it never came down below deafening. The slogans were very simple, 'Garh garh Singarh Singarh!', or 'Up up Singarh!'

What was truly remarkable were the teams of II and III termers who went around exhorting the company's cheering efforts to greater heights, till they were exhausted and replaced by the next pair of cheer leaders. The game was going fine with Singarh leading Cassino by 2-1. Then disaster struck and a Cassino company team cadet made a goal. He happened to be a Botswanian known for his superior athletic skills and silence. He had never been heard to speak... Singarh company contemplated in unison their own foreign team members nonchalance.

There were two foreign cadets on the Singarh company team, a Sri Lankan who claimed to be a national player and an Iraqi who did not claim much, but had arrived at the academy with an expensive hockey stick. The imported cadets were thus far not doing too much to allay the desperate Singarh cadets' hopes for a peaceful evening.

Defeat was not a happy option and neither was a draw. That only meant the team got away, not the cheering party which was just about the entire company. While these tense thoughts flashed in unison through the entire company's mind, the cheering was momentarily suspended along with normal thought. All that could be heard for all of three seconds, was only Cassino company's cheering.

The realisation that there had been no cheering for those crucial three seconds came as a rude shock to everyone affected. Singarh company's cheering resumed with even greater 'josh'...but it was three seconds too late. The company officers had turned around to see what the silence was 'not' about. A dirty look is worth a thousand words and the III termers too realised that their asses were on the line. The CUO and CSM also nicknamed the dragon and the wizard decided to lead the cheering personally.

Cassino Company redoubled its cheering and scored one more goal. A minute later halftime whistle blew. The Singarh II and III termers looked nervously at the officer's enclosure, the I termers looked to God to save their souls and water for their hoarse, burning throats. The company commander Maj. Ajay walked up to the team and patted them on their backs. The entire company tried to guess what might have been said but inspite of Maj. Ajay's smile the team captain did not look too happy. The company commander walked up to the company and said cheerfully, "I think the cheering was not up to the mark...maybe we'll do better in the next half...huh?!"

"YESSIR!" croaked the dragon and the wizard, the rest of the company just whimpered in support as it caught its breath.

That was enough, Singarh Company cheered like they meant it. A casual observer might have been excused for thinking the entire Singarh company's mothers were at stake. Then one of the Sri Lankan cadets scored a goal. A few minutes later the Iraqi saved a goal and Singarh company's honour. The company roared to its feet and had to be restrained by the drill instructors on duty. Even the officers jumped up and hugged each other. They defeated Cassino by one goal and Jay felt true happiness. His reasons were selfish, he just wanted to sleep peacefully. He didn't really understand the officer's joy.

They went onto the ground and hoisted the players on their shoulders. Singarh company would have taken a bullet for those two '*phoren* GCs' that day for having saved the day...and their night. The CSM and the CUO along with the rest of the cadets conveyed their congratulations and heartfelt gratitude. The foreign GCs of course couldn't understand the ado. The company commander asked the team what they would like as a reward and spontaneously granted them out pass before they actually asked for something he could not give. The Iraqi commented loudly he would have preferred a woman in his cabin to sleep with rather than out-pass to do the same. The rest just wanted a bed to sleep in. The company commander diplomatically ignored the Iraqi.

The foreign cadets were considered a pain for most part because most of them did not share the rest of the courses' hardship; hardships that they would have had to undergo at their own academies or country. They were quite physically fit...for most part. They were usually very good sportsmen and athletes and that seemed to be the overriding criteria for becoming an officer in their countries. At least the Indian GCs thought so, and wished they were foreign GCs. The rumour was the Sri Lankans just had to pass a running and long jump test to get selected to the army, they really were good at athletics. The Africans were said to be selected based on the results of a very, very long marathon. They were phenomenal cross country runners. Nobody seemed keen to jump the communication gap, which was quite broad.

In any military academy, intellectual deficiencies are forgiven if balanced by physical or sporting prowess. It is a historically accepted fact that those the Gods make dumb they compensate physically, and in the military they take first. That temporal issue would be debated later, but right now on earth and in the IMA there was always a place of pride and privilege for good sportsmen.

The I termers were culled for any athletic spark soon after their arrival. Most deliberately feigned ineptness...till they realised that sportsmen were excused all '*faltu*'[1] bullshit and fatigue. There was then a line of wannabes who beat down the CSMs door to try out for the company teams. Some genuine cases made it to the teams or were kept on reserve, but the pretenders came back worse for wear. The CSM did not take wastage of time, a precious commodity in the IMA, very kindly.

The focus of the season was boxing. Boxing was considered the game to build up a young soldier's character. It was also extremely physically demanding which made it kind of very dear to the academy curriculum. It did not hurt that it was considered fun to watch for most bloodthirsty types that normally make up any army or young male populace.

The inter-company boxing championship was coming up. Inter-company rivalry was not so much at the IMA where cadets were more concerned with survival and they were there for just a year and a half. The boxing championship went towards the overall champion company banner. The winning company commander's career would be enhanced. Thus the officers and other company instructors thus took the championship banner thing to heart. The cadets worked very hard to win as many competitions as they could as it enhanced their standing in the merit list which then meant a better chance of getting their choice of corps or regiment. But it's an immutable and somewhat unfortunate rule that there could only be one champion.

'Novices boxing' is an vital part of the academy's method of separating the wheat from the chaff. It was done through the same method-pounding in an enclosed space. It was also compulsory. Young upcoming urban professionals who now found themselves quaking in the military's grip were simply appalled. The thought of actually having to assault a fellow human without provocation was shocking to some, but eagerly looked forward to by other blood thirsty types. There were some who managed to hide the fact that they knew boxing. This lot actually managed to get through the novices boxing easily. They would fake amateur moves while winning ultimately with a well placed punch, thereby winning the instructors approval.

The rules of novices boxing were very simple. The Ist termers were paired weight wise and had to box for only one round of three minutes. Those three minutes were gruelling enough for a novice and many left the ring with shaky knees. That last could also be because of psychological rather than physical reasons.

Chib was one of those who believed what his mother said. She always said he was very smart. She might have exaggerated. He entered into a deal with Jay the night before the bouts to have just a mock fight. There would be heavy theatrics and dramatic blows with little actual impact or blows landed. They would somehow get through the three minutes without any actual damage to their faces or physique. There many such pacts made before the novices boxing.

There is an old army saying, 'don't teach your father how to fuck'. In keeping with the spirit of that saying, the academy through long experience had placed PT instructors including a DS to closely monitor the proceedings. The I termers were a bit taken aback to see so many instructors but took heart when they realised the officer on duty was Capt. Shah. He was relatively milder and the I termers hoped he would be kinder. They were wrong while DS Shah may not have believed in II and III termers handing out arbitrary punishments; he most definitely believed in awakening the 'killer' spirit in the cadets.

Jay and Chib began their bout as planned. A few seconds into the bout Capt. Shah said in his characteristic mild tone, "Box or I'll get someone else to partner you guys"

Jay and Chib turned around and saw the III term bantam weight boxing blue[2] contender warming up in the background. This was a stocky, totally muscled to the eyebrows Neanderthal from Sports School Rai[3]. This boxing bull had been trained from the tender age of 12 in the pugilistic arts and had fully imbibed the 'killer spirit'. Chib looked into the bull's eyes and got a full dose of the killer gaze. Chib suddenly re-discovered his killing spirit and launched himself with a fury that took Jay by surprise. Jay went into a clinch and managed to whisper urgently, "What the fuck are you doing man? We had a bloody deal!" before they were pulled apart violently by the referee.

It all came to an end when Jay managed to land a solid punch by sheer chance. Chib promptly sat down covering his head and refused to get up. Jay was declared the winner and the last Jay saw of Chib took away the joy of victory...somewhat. Capt Shah was forcing Chib to 'stand up like a 'man', the raging bull was invited in to give a fresh infusion to Chib's anaemic killing spirit. Chib face was too swollen for any recrimination that evening when Jay dropped by. Jay brought him some brandy in a glass with a straw and some painkillers to take away the sting of defeat and excess infusion of killing spirit. Chib's surviving eye looked at Jay accusingly and he croaked something that sounded like 'auck you basshard'.

Jay was one of those selected for boxing, a sport he had only experienced in not so friendly encounters with his sister. Jay was chosen not because of his boxing talent but because of his weight. The main fighter was an ambitious II termer who needed someone to be the mandatory reserve. Being part of the boxing team was however a novel experience, one was excused punishment parades and other inauspicious gatherings. The downside was the weight loss or gain regime that included dehydration by being wrapped in carpets or being stuffed with bananas and milk. But it was still all good, the gladiator like feeling. They who were about to die were grateful in this case.

GC Rohan Ranatunga was a six foot two inch hunk of prime Sri Lankan beef. He also claimed to be some kind of amateur national boxing champion. With his rippling muscles, lantern jaw and broken nose he looked the part. Singarh Company welcomed him gratefully into the boxing team with sigh of relief. The heavy weight category in the academy is hard to populate, with most cadets being 60-70 kilos on an average. In any case those heavier were whittled down to that uniform size after a few months training at the IMA. The 60-70 kg weights came under the bantam or welter weight categories and were very fiercely fought, having the most talent. Peres being a heavy weight champion was excused the gruelling training regime of running, sparring and practise bouts.

The practice bouts with serious senior boxers were the scariest. These were experienced disciples of the game who took boxing very seriously, almost religiously. They wore thin punching pads instead of gloves and could inflict serious damage if they thought you were shamming while sparring with them. Jay had managed to impress them with his aggressive spirit and was let off with just a light hammering.

Rohan the champion roamed around blithely giving unwanted, but graciously accepted advice, especially to the impressionable I termers. He enjoyed the special diet and even the special masseur hired by the cadets. The only other cadet in his weight category was a burly Sikh Ist termer from Alamein Company. This sardar or surdy as he was commonly called, had been fed and raised lovingly by his doting mother. And this extensive dose of dairy products had ensured he reached this current XXL size. Mummy*ji* back in Bhatinda would have been horrified to learn her darling fatted calf was being readied for slaughter by Rohan Ranatunga the butcher of Singarh company nee Sri Lanka.

The fatted calf of Punjab was also terrified. He had heard the rumours of Ranatunga's physical attributes and boxing talent. He had begged the CSM to excuse him from the team. The CSM sympathised, but was helpless. They needed every point they could get and a RSC (referee stops contest) got more points than a walkover. The fat surdy was very aggrieved at being sacrificed for just one measly point. He was willing to die for his country, not for just one lousy point for his company. But that point could make the difference in the boxing championship ranking. So surdy boy bit the bullet and munched his special extra diet. He was not overly burdened with training by the senior GCs in charge of the team. After all he was expected to last just half a round before the referee stopped the contest on humanitarian grounds or it ended on a knock out. 'At least' he thought to himself it would be over in one round and he would not feel much pain after being knocked out. He made plans for the few days SIQ he would get from the doctor after getting mauled.

The great day of the boxing finals dawned. The Ist termers cheered lustily for their company fighter. This was based on their previous experience during the hockey match. They were mistaken. The rules for cheering during a boxing match were different. The first minute after they started cheering the PTO (physical training officer in charge of the academy) who was also the referee for the match stopped the bout. He called the BCA and said something. The BCA snapped to attention and turned red. Red is always a sign of danger whether on traffic lights or someone's cheeks. The BCA called his sidekicks and they in turn went up to their respective I termers and told them forcefully to cheer only during the breaks in the bouts.

The sight of blood and violence can bring out a certain excitement and misplaced exuberance in men. A few I termers who still cheered during the bout were quietly culled from the herd and led to the side lines for a spot of warming up. The rest got the message and cheered vigorously but only during the breaks.

Then the bout everyone was waiting for...Rohan versus the fat surdy boy.

Rohan entered the ring, did a victory lap and waved his hands at the crowd. The cadets went wild and cheered loudly and even the officers did not mind the spectacle. Emboldened, Peres even bowed theatrically to the commandant, which brought further applause from the entire academy. There was some raucous whistling which was dealt with summarily and died out abruptly.

Rohan then went back to his corner where his seconds massaged his shoulders as he smiled compassionately at his opponent, even winked encouragement. His opponent was only thinking of dear mummy*ji* and how the pain and embarrassment would soon be over. The bell rang and Rohan danced towards his opponent. Surdy closed his eyes and decided to land at least one punch before he was mauled by Rohan the champion. The punch landed and after a while surdy heard the referee counting. He supposed he had been knocked out. He opened his eyes lowered his guard and was surprised he was still vertical. Puzzled, he looked around for his opponent. The scene was not what he had expected. Rohan was laid out and refusing to get up. The referee counted Rohan out and looked at the judges. Before he could make any pronouncements the commandant said grimly, "That man is bloody shamming! The fight will continue!"

Rohan was hustled to his corner where his seconds looked positively disappointed and sulky. A doctor quickly checked Rohan out of medical and international considerations and declared fit. The surdy on the other hand was now more terrified of the commandant's wrath than Rohan's revenge.

The next bout started off slowly with both boxers displaying a marked reluctance. There was no cheering only suppressed laughter and giggles. The commandant looked at the Alamein and Singarh company commanders. They being old soldiers got the message and immediately went up to the ring and said 'fight you cowards or else...' or words to that effect. Those words had no effect on Rohan who was concentrating on saving his already bent nose from further realignment. The surdy from Alamein however was more terrified of the company commander.

The round continued and degenerated into the ridiculous, with a fat sardar chasing an unwilling Rohan around the ring. The surdy was pausing for breath every so often while the Sri Lankan who was fitter was able to stay just out of reach. By the third round all the officers including the commandant were smiling. The only ones not laughing too hard were Singarh company officers, NCOs and cadets. Rohan and his 'Gandhian' approach to boxing would most likely cost them the championship.

The bout finally got over and not too soon. The Alamein company boxer was declared the winner and not a minute too soon. The surdy was sweating from every pore and hole and could not have sustained his pursuit of Rohan around the ring for one more minute. In fact the Alamein company commander had been worried his chubby amateur would drop out of exhaustion and Rohan would win by default. The sardar from Alamein won and raised his gloved hands to the sky and silently thanked '*Wahe Guru*' for saving his sorry ass from getting whipped as planned by the rest of his team.

The seconds and the rest of Alamein Company rushed to carry the 'surdy' now respectfully called'Singh sahib the great' away on their shoulders. Rohan was seen off by dirty looks and even his seconds and Sri Lankan mates averted their faces.

That night Singarh slept uneasily wondering when they would hear the announcement that would make the wet PT ground their bed. But the Singarh company commander was having one of his humane moments. He had decided to not to screw everyone for the cowardice of one and a foreign object at that. He had already decided on the appropriate and more satisfying course of action.

The next day a Sri Lankan officer, the defence attaché was seen observing the drill square routine. He then had the Sri Lankan cadets separated and personally ensured they were drilled to an inch of their lives. That evening they heard Rohan grunting his way through rigorous bruising exercises in triple pack. The other Sri Lankan cadets were also there and a lot of harsh Sinhalese words, most likely foul, reverberated on the PT ground. The disgrace of Sri Lanka in the boxing ring had been felt keenly by the Sri Lankan military attaché. He had decided to come down personally and set matters right. He had been sent a video of the entire bout by an officer in IMA. The cadet appointments felt vindicated though beaten to the punch. The other GCs felt relief but no sympathy. There was no trouble after that with any of the foreign GCs. It was better to shed sweat than blood on some lonely foreign field...especially PT field at that.

NOTES

1. Faltu, unnecessary
2. Blue—indicates a high level of proficiency in a sport or discipline. 'Blazer' is the highest and this system of grading was started in Eton and continues to this day.
3. A very famous school in Haryana India that has produced many fine sportsmen and military personnel. The thrust in the school is on ensuring all students excel at one or more sport disciplines along with academics.

Chapter 15

THE CLASS ACT

Most students could only dream about going to classes in beautiful outdoor settings, with the sun shining brightly, butterflies and singing birds. The average IMA cadet usually slept in class and dreamt of food. The sunshine had the same effect as a lullaby. Hell! Even the instructors barking sounded like a soothing invitation to sleep-not that any invitation was needed. Not that the instructors ever stopped...

An unknown cadet once remarked, "After joining the bloody academy I get aroused looking at food and hungry when I see a girl!"

After a few weeks in the academy Jay felt like a zombie. Arousal and hunger would get submerged in the dreamless comatose state that often overcame cadets at all times in all positions and situations. It was fondly mistaken for sleep when they got home. There were some classes where sleep was out of the question, like PT or drill obviously. Outdoor classes were also not always the best of places to practise the art of shuteye. It was very exposed with good lighting and so on. There were also just too many landmarks around which the cadets could be sent.

This last was done for reasons of man management; it wasn't easy holding the attention of so many young men. Only 1% of the class took some interest, these social pariahs were nicknamed 'Keen Kumars'. These sincere souls actually listened to what the instructors said, even took notes. These *chutiyas,* as they were fondly termed, even fought for the dubious privilege of sitting in the front row.

The rapt attention of a class can be very gratifying. The visible disinterest in the form of snoring or even sleeping in bunches right under the instructors nose can be very insulting and discouraging. Any public speaker who is not a politician will tell you that.

When the insulted teacher is an unhappy instructor things tend to go South. Sometimes senior officers dropped in to see how the lecture was going. These were the tricky moments, with some predictable outcomes. An avuncular visiting senior officer may gently nudge a cadet awake. Some just smiled indulgently and ignored the snores. But there were the sticky ones, who simply pointed out the sleeping beauties to the instructor. This implied a slur on the instructor's instructional ability. The sight of an instructor waffling on without anyone paying attention, with some actually sleeping, indicated lack of class control. This always upset the instructors especially if their bosses were observing.

The situation was usually ameliorated by sending the class for a quick 'round the round' of the nearest or furthest landmark. That depended on various factors. These factors were the affability of the instructor, weather conditions or time of the day (everyone felt sleepy in summer after lunch), and of course the seniority of the visiting observer. A strong rumour circulated among the cadets that the viciousness of the instructor also depended on the phases of the moon and or his wife's menstrual cycle. These hypotheses have never been proved.

Today they had a 'LecDemo' or lecture demonstration, also known as 'LDMs'. Jay was actually looking forward to it. The LDMs were more fun...well maybe not so much fun as interesting. There were lots of bicat strips[1] and grenade 90s exploding. The demo (demonstration) area, during the course of the lecture cum demo have smoke swirling around and men in camouflage taking positions charging mock machine gun posts, even dying dramatically. All this was lapped up by the Jay and the other I termers. Some of them even clapped and whistled during their first LecDem. That had been a mistake. The commandant had been one of the observers and he had not liked what he had observed.

"Unofficerlike and bloody ungentlemanly!" The commandant muttered, "Behaving like this is some *tamasha* in the bloody bazaar!" And then he had stalked off with his entourage.

That had been enough to drive the instructors into foaming frenzy that lasted an hour. There were many bruised backs and scraped knees and elbows by the time the class was over. The knowledge of the students had been enhanced and they had learnt a valuable lesson, though not in the syllabus. The experience had subdued the enthusiasm for applause and exhibiting any vocal approval of military instructional techniques or *tamashas.*

Now the cadets were sitting demurely, waiting for the only interesting part of the day. The only part that made them believe they were actually on the path to becoming a warrior. You will appreciate it's difficult to feel like a warrior while being abused and yelled at all the time. The fact they were all wearing uniforms stained and torn from rolling or crawling did not add to any martial feeling or fervour. But at the LecDemo they heard military stuff, actually saw stuff, which they might have to do to save the country. It was good stuff by any account. But even so there was always some idiot who went to sleep and spoilt the smooth run. It was usually the foreign cadets, but in the spirit of bilateral relations the instructors would invariably screw the entire class.

Jay looked apprehensively at the group sitting next to them wearing II term badges and I term looks. The technical course or techos as they were known went directly to the II term, in honour of their having slogged it out at some engineering or technical college, for a year more than the other graduates in IMA. They were also newcomers, but at a distinct disadvantage as compared to the I termers as they got no leeway. Being newcomers and usually a bit above the average cadet in intelligence, the techos tended to screw up things in more complicated ways. In any organization, complicated screw ups by juniors meant frustration and above average aggravation for seniors. This was a situation to be avoided at all times. But the techos took some time to absorb that, usually when it was too late. Jay's apprehension was merited and he voiced his grave concerns to Kapoor, "I hope those techos are not going to ask any *chutiya* type questions and get us all screwed!"

Kapoor looked over at them, some techos were already busy scribbling. He shook his head and looked at Jay sadly, "No chance...we are going to get screwed again...may as well take a nap!"

The techos being an intelligent lot as mentioned earlier, were also comparatively nerdier than the average male cadet and tended to ask vaguely intelligent questions... These questions were of the logically uncomfortable variety. A prime example would be the question being asked with the next few minutes by a techo at the LDM.

"Sir! Why do officers have to lead from the front? He will be the first to get killed!"

The conducting officer smiling jauntily while clenching buttocks and teeth answered this logical question, "Well... officers always lead from the front...it's what we're supposed to do!" He also had doubts privately but...

"Is it because of tradition or necessity sir?"

"Well that's what officers are there for!" this through very gritted teeth.

"But sir..."

By then the entire course was clenching buttocks, teeth and fists. It happens when you find yourself in a vehicle being driven deliberately towards doom.

By then the officer had had enough. The discourse was heading to the stupidity of war and the absolute idiocy of killing and ultimately unemployment of armies. What then?! The thought was terrifying. Better to just lead from the front and hope to survive. "Ok enough! You meet me after the class! Anybody else with silly bloody doubts...anyone? Gentlemen...? "

This last word could be condensed into one smaller word-'shutup'! Jay and about two hundred I termers tried telepathically to communicate those six wise alphabets to that techo but it was too late.

'Meet me after the class', was a thinly veiled threat to the pesky student. But some, especially the pesky techos took the statement seriously and accepted it at face value.

Thus it came to pass that the poor bloody techo actually went to meet the instructor during the break to clear his doubt. It was what they called a 'genuine doubt' in military parlance. After all why should the officer wearing all his badges of ranks lead from the front where he would be killed? The army had still not learnt its lesson after the I world war when officers were usually the first to be killed. It was also true that the stupid and the innocent got eliminated early. Look at the dodo bird. The techo did not go the way of the dodo but you will find him at the end of this story wishing he would.

The dodo reborn as techo stood silently near the officer's enclosure during the break waiting for the instructor to notice him. He was duly noticed and ignored by the instructor. Then one of the senior officers there as observer felt sorry for the sincere techo and beckoned him over. After all it was the rare cadet who had forgoes his tea break and snooze in the quest for knowledge.

"Yeah son what are you looking for?"

"Sir, that sir told me to meet him after the class" said the dodo earnestly pointing at his nemesis

The other officers there sniggered at this fatal sincerity. Here was someone who was not sure about leading from the front but was rushing headlong into doom. The exasperated instructor who was dodo's target tried to shake off the embarrassment by snapping, "I said after the bloody class not during the goddamn break!"

"Where should I meet you sir?" asked the techo.

The other officers were now openly guffawing. The instructor saw red as his face turned the same colour and said, "Come to my office in undies after the class!" That was a clear signal to leave in the face of imminent death. The techo had still not read the writing on the wall and asked, "Sir do I need to bring any writing material...to take notes..."

The other officers including the senior observer were laughing their heads off. The instructor paused for a long time and the said "Nope! You seem to want a fingering...I think your fingers will do just fine!"

The drama had been witnessed by Jay and the entire Ist term course. They watched apprehensively, while the rest of the techos watched approvingly. They were more or less on the same grid. Their thought process was still not tuned into the IMA or military frequency. For some techos the 'tuning' in to the military grid would never happen throughout their service.

Jay and the other Ist termers had tuned in and were thus concerned with getting to the next class in time. 'Being in the right place at the right time in the right dress will take you very far in the army'. This was an old military axiom that held deep meaning for the cadets-it was essentially a survival mantra. The next class was also an 'outdoor', on a very interesting subject-judging distance. The subject was interesting and supposedly fun but not their instructor.

This was not a combined class for the entire Ist term, just the Ist termers of Bhagat battalion. The instructor was a Capt. Cheema from the famous Guards regiment[2]. He was a gallantry award winner, athlete, tall, dashing and dumb. He was also totally callous when it came to cadets. Capt. Cheema considered cadets animals that needed to be physically fit and not really in need of any mental stimulation. After all brains were not what had gotten him this far.

The Ist termers looked at the vision that Capt. Cheema presented as he leaned casually against his Yamaha 350. The sun glinted off the 16 brass buttons on his Guards SD (service dress), missile badge and aviator sunglasses. He had deliberately worn the SD rather than change into a camouflage field dress... that would have not had the same impact.

The Ist termers had mixed feelings—also called 'love hate' relationship. The Ist termers were impressed and depressed at the same time. Impressed with Cheema's looks and depressed by the thought there was such a big SOB inside the beautiful exterior. It was obvious that said bastard would teach them a lesson rather than just teaching them.

Cheema opened up the proceedings by saying, "You fuckers are late, get on your hands down!" It wasn't true but there was no system of appeal.

After 100 pushups the Ist termers were ready and thoroughly warmed up. Capt. Cheema was a firm believer in warming up. Warming is essential before any physical exertions but generally accepted as time tested method of getting the class's undivided attention. The Ist termers were all anxious ears and burning arms.

"Right then can anyone tell me what are the methods for judging distance?" asked master Cheema.

There was an anxious silence as everyone racked their brains while looking nervously at each other. Cheema took the moments of uncomfortable silence to look for a suitable landmark around which the cadets could be sent for a bracing jog. 'Jogging jogged the intellect!' or so Cheema was fond of saying. He was proud of that witticism...it being the one of the few passable or printable ones his intellect produced.

No answer seemed to be coming forthcoming. Some of the Ist termers had begun loosening their leg muscles in anticipation of the highly imminent order to sprint.

"Sir! Usha ki bra!" cried out a Ist termer desperately breaking the desperate silence. The entire class looked at the wise guy who had given a silly answer but also mentioned a bra. Surely something the DS would not forgive...mentioning ladies underclothing in public. All eyes were on instructor Cheema who looked inscrutable in his aviators. The seconds ticked off, and then Capt. Cheema grunted "Yep! That's right USHA KI BRA is the acronym for the methods of judging distance." The entire class drew a relieved breath, as the Capt. proceeded to expand on what the acronym stood for.

USHA KEY BRA was mnemonic that stood for Unit of measure, section average, halving, bracketing, appearance etc. etc. Classes were supposed to be prepared for in advance and the instructor was there to just clear doubts. But alas most of the gentlemen were there at the academy...because they thought they would not have to study any more. True, this was not rocket science but it felt that way when hungry, exhausted and of course sleepy in the extreme. For every smart alec who had prepared, there were ten sleepy dunces who could not answer the instructor correctly. Thus by the end of the first half hour, half the class was on its hands down.

The educating part continued for another fifteen minutes and then Capt. Cheema began to get bored. He gazed once more at the horizon as he polished his sunglasses. There was an air of finality about his actions. He then popped the question everyone was dreading, "What's the distance of the lone tree in the middle ground?"

This was not a trick question because it was part of the tradition. No answer was the right onc cxccpt thc instructors...that was also part of the tradition, as was the next instruction from the instructor. No one thought it strange that no cadet had given the right answer in the last 70 years or so. The cursed lone tree had also been there forever. Jay swore like others before him, he would come back someday and cut it down.

"Fuck off... all of you—go and measure the distance...on foot!" he added as if there was nay other option, "... the first five will stay the rest will repeat"

There was mad rush as the cadets dropped their satchels stumbled over each other's water bottles, stepped on those who had dozed off or were slow to react. Capt. Cheema smiled as he slowly took off his glasses so he could watch the Ist termers raise the dust in natural tones. He took the opportunity to scratch the stump inside his prosthetic lower leg and thought about the time he had run around the same tree just eight years back. 'God!' He mused like countless before him, 'that tree never seemed to die inspite of all the curses and ill will heaped on it in the last 70 odd years'.

He wished he could be there running with the cadets but an LTTE booby trap had finished all that. He was just lucky it was the right leg and the Yamaha 350 had its gear lever on the left side. 'Pity' he thought since he was fond of the Enfield which had its gear on the right side. Capt.Cheema smiled at the sight of the cadets raising dust. He decided he had had enough fun, prolonging it would be just cruel. He was fond of the cadets in his own way, so hopped onto his bike and left.

The sweaty dusty horde came back and fell on its knees out of exhaustion and gratitude to providence for having removed that devil Cheema from their midst. Their joy got over in just two minutes. The outdoor classes were over, the Ist termers rushed from the stands for their next class on military writing. The outdoor classes were in the Tons[3] riverbed and the indoor classes were across the other end of the academy. They needed to really hurry if they wanted to get there in the fifteen minutes. A momentary pang of pity was spared for those poor souls on foot as they struggled to form squads and slowly shuffled off in a 'limp trot'.

The 'limp trot' not to be confused with jogtrot—since many of the cadets had some injury or muscle soreness thus reduced to a slow uneven jog with a limp at every alternate step. They actually overhauled those on cycles since the entire route was uphill and at places steep gradients that forced the cyclists to dismount and drag their unwilling cycles.

They finally got to their class and saw a decidedly academic looking officer standing impatiently outside. He was wearing AEC (Army Education Corps) insignia and a pout. The Ist termers were wary. They had heard of Maj Sharma earlier and were not cheered up.

The army education corps is a very mild organization with little or no expectation of bullets ever flying their way. They would almost never see combat unless ambushed on the way to bed by a frisky wife. Some however nursed secret martial ambitions and liked to unleash the full splendour of their rank and position as instructor. Sharma took his time till all the Ist termers of Bhagat battalion had formed up. Some were still retching and heaving from the exertion of a dead sprint the last 500 yards. He adjusted his glasses, and then decided to take them off and polish them as menacingly as he could. Jay did not however feel the menace only acute disgust that they might have to suffer at the hands of this uninspiring insect.

The standard starters of pushups were ordered, and then they were ordered inside. It was past noon and hunger pangs were becoming a serious distraction. What was downright disturbing was the drowsiness. The academy etiquette and orders on the subject of drowsiness on duty was very clear. In case a cadet felt the need to sleep was overwhelming, said individual should remove himself to the rear of the class and stand. In this case the action could be carried out without asking for permission, disturbing other sleepers or disrupting the class in any other way. It was also the duty of every sleepers neighbour to keep him awake. In the event of a cadet being found sleeping all immediate neighbours were liable for punishment. In the army you were your brother's keeper.

The sleepiness crept slowly over the person till dream and reality merged and one only woke up when the instructor fired a well-aimed piece of chalk at the sleeper. The instructor took this task very seriously. He considered it a personal affront and immediately ordered the sleeper and his half asleep neighbours to the back of the class and assumes the high push up position aka legs high up on the wall and hands on the floor. It was a comfortable position to grab a few minutes of sleep before the blood started bothering the brain.

Halfway through the period almost 80% the class was either standing or in push up position at the rear of the class. There were only four cadets sitting with intent looks on their faces. They were those who had attained the highest level of meditation and could now sleep with their eyes open. Jay was not one of them.

The instructor was at a loss. Then he remembered what one of the other instructors had said about dealing with situations like that. He told the entire class to go around the block and get back. But alas he had not specified when the cadets should be back. The more enterprising dove into the bushes and went to sleep. Sharma felt his army had thinned out somewhat and a head count confirmed the same. He then sent some Jay and Kapoor to look for the others while he went inside to polish his glasses. Now the searchers joined the other snoozers in the bushes.

The siren rang just then and before Sharma could rush outside, the Ist termers were a distant dust cloud. He considered complaining to the Bhagat battalion staff officer but they might laugh at him. He decided he would just make up some story about how well he handled the cadets. The others might not believe him but the Education Corps comrades would. Admiration of the fairer sex is gratifying, but that of peers very uplifting. He would meet these cadets again and then he would sort them out.

This story did not have a very happy ending. The techo was later seen wearing full battle dress writing 'Officers always lead from the front' on the dust in front of the company office. The instructor had meant it when he had said writing instruments were not required. The techo was writing with his finger. He had vaguely accepted he would be 'fingered' by the instructor just not this way. Jay and the others gave him a 'thumbs up' for encouragement as they passed by. The techo showed them the finger. He would not be questioning military logic or tradition for the next twenty years at least.

NOTES

1. Bicat strips are crackers set to explode in intervals to simulate small arms fire, grenade 90 were used to simulate the sounds of explosives/grenades.
2. The Guards regiment or The Brigade of the Guards as they were called had been formed out of some of the oldest infantry battalions of various regiments. Till recently they recruited soldiers of above average height and as a consequence had officers who were usually taller. The uniform of the guards regiment was also flashier than that of other infantry regiments.
3. Tons is an actual river at the bottom of IMA, and the riverbed is a wide open space which has most of the outdoor classes.

Chapter 16

MANIC MONDAY PART 2—THE DRILL SQUARE ROUNDS

The orderly knocked at the doors of the cadets. Jay waited for the pause and then heard the thump clash. It was a Sunday and some of the Ist termers were depressed. Jay opened the door and looked at what the orderlies had left a highly glossy pair of ammunition boots[1] that looked like it had been made out of a solid block of black metal. Kapoor too opened his door and looked Jay. They both looked at each other and the boots sadly.

The boots were a combination of leather that felt like wood and metal. The thump had been the leather part and the clash that of the thirteen regulation steel hobnails in the sole. The iron toe cap and unyielding aspect of the boot probably accounted for the drill instructors personality too. Jay pondered wearily on the neurological effects of bone jarring foot stamping...

Many generations on the drill square had pondered similar issues before succumbing to the mind numbing effects of high decibel commands, exhaustion and foot pounding. The idea was an old one, to train a person to act automatically. The body would act even if the spirit had fled. The drill square may 'fuck and forget'[2] you, but the drill square tests or DST as they were popularly referred to left long term psychological scars. A I termer once asked innocently if PTSD (post-traumatic stress disorder) was something similar to DST. Jay could actually believe that now.

The DST was a test conducted every term and passing was mandatory for graduating to the next term. It was also highly advisable to pass the DST in as few attempts as possible. The drill instructors were responsible for all formal discipline and punishments in the academy. They were also rated on the number of Ist attempt DST qualifiers they produced. This was one of the reasons for Jay's depression. They were having extra classes with the CSM that morning as they had the DST third chance on Monday.

The other and more serious reason for Jays depression on this bright sunny Sunday morning was the out pass factor. Not being permitted to go on out-pass was one of the unfortunate fallout of being a DST failure. In Ist term the humiliation was further rammed in by being forced to wear a DST number on your chest at all times, which made one look and feel very much like a concentration camp inmate.

Bhasin the BCAs understudy and resident quisling of the Ist term walked by with a smirk. "We are off on out-pass what about you guys?" asked Bhasin cheerily.

The fact that he knew why they could not go made the question even more irritating as Bhasin meant it to.

Kapoor fell into the trap and said, "No man we can't go we still haven't passed the frigging DST!"

And that gave Bhasin his supremely insufferable moment which he seized with both hands and said with infinite superiority, "I know..."

He had not known, however, that it would have earned him two swift kicks from Kapoor and Jay. It made them feel better for the next five minutes before despair hit them again. They went back to Jay's cabin and wondered what to do. They steeled up their nerves quietly for a few minutes as they pondered the unfairness of the cosmic drama. A drama that allowed reptiles like Bhasin to pass the DST and go on out-pass while the good guys suffered on Sundays.

It took another ten minutes before Jay and Kapoor came to the same conclusion. Kapoor went first and said, "Fuck the DST, fuck the IMA and fuck Bhasin! I'm not going for the DST practice."

Jay considered the proposal gravely and seconded it, "To hell with everything, lets skip and go to town!"

Kapoor agreed with Jay in principle while they both tried to figure out the mechanics of sneaking out of the academy. They saw one of the II termers walk past in decidedly civilian clothing. He was wearing a *kurta pyjama* and a turban. The cadet looked confidently civilian and very fit considering he was SIQ (sick in quarters) and also due for punishment parade. They followed him to the hedge next to the IMA perimeter where he vanished.

"That's the way we'll do it in civilian clothes!" said Kapoor jubilantly. "Nobody will be able to catch us..."

They went back to their cabins and picked out the most civilian dress they could find-jeans,T-shirts and Nikes. A lump came to Jay's throat as he beheld the vision in the mirror. The clothes were two sizes two big and the reflection was that of wild eyed desperado. They didn't care; they were going out into the normal world and would perhaps look and feel human for some time. However, in the excitement of breaking academy rules in their quest for transient freedom they forgot a very important thing.

A GC with his distinctive crew cut hairstyle is a well-known sight in the Dehradun. Civilian clothes or makeup could not hide their true identity. Cadets on bona-fide outpass were supposed to wear regulation grey lounge suits in winters and grey trousers with white shirts, brogue shoes and the IMA tie. They were also forbidden from using public transport; they had to use the IMA issue bicycles. There were drill instructors patrolling the streets looking for cadets 'improperly turned out' and of course the few who felt they could outsmart the system.

Kapoor had wanted to use the IMA cycles but had been dissuaded strongly by Jay so they commenced on foot. The gap in the hedge led to a hole in the academy perimeter wall next to the main highway. They walked rather gingerly the few meters to the bus stand. It did feel somewhat like covering open ground in front of a hostile machine gun bunker. They waited for some suitable vehicle which would give them a lift into town. A couple of vehicle did slow down but sped off the moment the drivers got a look at them. They did not look ideal candidates for getting a lift given their ill-fitting clothes and lean mean and hungry looks. Such looks repel friends and foes alike.

They had waited ten minutes before the excitement of the morning finally got to Kapoor and he went behind the bushes to pee. Jay waited patiently whistling cheerily as he imagined the juicy butter, *tandoori* and other types of chicken he could ravish while downing glasses of thick cold *lassi*. He was already kind of aroused when he saw something which deflated him instantly. A burly drill instructor was slowly peddling up the slope. He was huffing and puffing from the exertion and was bound to blow Jay and Kapoor's dreams down.

Jay decided to follow the lesson he had learnt on camouflage and concealment, by staying very still and hoping the ill timed cloud would blow past. He then heard Kapoor whistle as he shook himself happily and finished up. Standing in the shadows of the bus stand Jay was partially concealed and it was just possible he would be ignored. But Kapoor would soon step out of the bushes and be brutally exposed to the lethal gaze of the drill instructor. Jay tried to warn him by hissing, "Careful fucker there's a drill *ustaad* coming!"

That only made things worse. Kapoor came out and addressed Jay directly, "Speak up *saale* I can't bloody hear you."

The drill instructors were not the brightest bulbs in the IMA chandelier of instructors; but even the dimmest wit would have recognized two things instantly. Firstly the guy with the crew cut, tan line on the forehead because of the regulation beret had to be a GC. Secondly the guy he was talking to so cheerfully had to be a friend, ergo another GC.

Jay and Kapoor came to similar conclusions about the drill instructors deductions and were faced with the classic choice-'fight or flight'. Fighting was not an option; flight was something the last few months had toned them for and flee they did, very very swiftly indeed. They managed to find the hole in the wall and were back in their platoon in a trice.

They went to their room and changed quickly into their drill practice dress and managed to get to the fall in front of the CSMs room just in time. Their fat nemesis was just a few steps behind them. He went up to the CSM and whispered something to him. The CSM looked directly at the latecomers Jay and Kapoor. It felt like a very long time and then the CSM shook his head. The drill instructor clumped off in his heavy drill boots sounding like a disappointed robot.

"You two on your hands down and give me hundred!" was all the CSM said. Jay and Kapoor complied gratefully. The Sunday adventure could have ended worse. They practised the DST routine long after everyone had broken off. This was one test they had to pass....

It was Monday and a very important Monday (which one wasn't). It was the day when they would along with almost half the Ist term course from all the companies would be reappearing for the DST. It would a miserable life full of extra drill, a fate worse than death, if they failed, again.

They were lined up steaming in the cold after having been suitably warmed up. This day nobody minded the warming up on the drill square. Anything that would help the DST hopefuls go through their routine with the grace and fluidity to impress the IMA adjutant would be accepted gratefully.

The drill instructors lined the cadets up giving out words of caution and encouragement. They were the only friends and well-wishers the Ist termers had this day. The drill instructors put the final touches on their uniforms straightening up term badges, academy flashes and shaping berets correctly. A proper warming up on the drill square can leave a lot askew mentally and physically.

The Drill Subedhar Major (SM *sahib* as he was plainly called) came on the scene He was a majestic looking figure with huge mutton chop whiskers, with fierce eyes and a hooked nose. He looked for all the world like a cross between a lion and an eagle. The academy drill SMs were usually from the Grenadier or Guards regiment. This gentleman was from the Grenadier regiment and a decorated soldier, due to retire in a year.

There was an instant hush as he came closer with his pacing stick[3]. The senior drill JCO at the scene bawled out '*sawdhan*' (attention) and the entire DST parade slammed their feet into attention out of awe, fear and other such emotions. The sound of a hundred boots on the tarmac was sweet music to the drill SMs ears and he smiled slightly or at least his whiskers quivered briefly.

He said "*Shabash* GC!" The other drill instructors relaxed while the cadets stiffened up. The next part of the DST, the dress inspection would now commence. The Drill SM walked slowly among the ranks of cadets his keen eyes missing nothing. For all his scary personality the drill SM was a compassionate man. He had seen many cadets and also lost a son who had joined his regiment to insurgents in Manipur. He still had a job to do and he could give very little quarter.

There was one stupid Ist termer who had no bothered to get his boots waxed and was booted off the line. There were a few more with twisted laces and missing buttons but were left off with a little more warming up. The cadets were asked to raise their left legs and then their right. Their ammunition boots were supposed to have thirteen regulation hobnails with the socks being folded just so-in the regulation manner. The inspection went off peacefully without too much blood or sweat. And they waited for the next demon of the drill square-the IMA adjutant.

The IMA adjutant trotted up on his horse with gleaming riding boots, eyes hooded by the visor of his peaked cap, gold braid and brass glinting. The drill instructors stiffened up, the cadets froze and the SM screamed, '*saaaaawdhan*'. There was another clash of metal on tarmac and the adjutant and his horse both nodded their approval. The horse even made appreciative noises.

The adjutant's horse like the adjutant and the drill SM was an institution. It was fairly old and very staid. He had seen many adjutants and cadets and was used almost exclusively by adjutants because he was used to the drill square. Most importantly could be relied upon to stay still.

The last time a dashing polo playing adjutant had decided to come onto the drill square on his favourite polo pony, it had ended in embarrassment. The poor animal had been so spooked by the sight and sound of so many humans screaming, crashing weapons and stamping the ground, it did the only thing polo ponies are meant to do-run. It had been the subject of much mirth and much embarrassment for the office of the adjutant. Even the drill instructors had briefly and furtively smiled at the sight of the adjutant desperately trying to control a horse intent on jumping over the hedge.

There was also much speculation that the adjutants horse actually decided the pass or fail status of the DST candidates. Indeed it did bob its head up and down and sideways after each cadet finished the DST routine. There was pregnant pause while the cadet would wait for the verdict. The horse would snort and bob his head, but the adjutant did not always agree with the horse trusting more in the SMs judgement. He would look at the SM and nod imperceptibly who would then yell out the verdict. Only the SM understood the meaning of that nod. One cadet was declared passed, but then promptly failed on the spot for doing a jubilant fist pump.

Finally came Jay's turn. He was sweating from the tension and the waiting inspite of the cold. He understood how it felt waiting for the firing squad. The warming up had gone cold and he felt stiff. The drill instructor realised that and immediately ordered flatfoots[4] for the entire waiting detail. For once the cadets accepted the order with gratitude.

Jay began his routine, made few mistakes and invoked God silently for mercy. God's representative on the drill square aka the adjutant looked emotionlessly at Jay and then nodded to the SM who simply said "chest number 102 pass"

Jay controlled his emotions and joined the ranks of those who had passed. Kapoor was one of those who had passed. Kapoor had actually gone and fed the adjutants horse apples surreptiously. The bribe seemed to have worked and nobody could convince Kapoor otherwise.

The cadets who had passed avoided the accusing looks of those who had failed. They felt righteously aggrieved with a good reason. They had after all done the routine as asked but still failed. They felt betrayed the same as any man abandoned on the battlefield by his comrades.

The Gods of drill—that is the adjutant and the SM were just looking for some potential. Any GC who gave even the slightest indication of having absorbed the basic idea of drill-timing, posture and some semblance of grace, would be passed. But then of course the adjutant could be in bad mood and then pass or fail could very well be decided by the horse. Today he was in a good mood but he still could not pass everybody-it would have been bad for his image. Jay and Kapoor were not sure if they had any potential or any image. They just felt gratitude but again not sure to who or what...God, The adjutant or his horse. They would toast the holy trinity of God, horse and adjutant over *lassi* sometime soon when they went on outpass.

NOTES

1. Ammunition boots have more or less been phased out but still used exclusively for drill and mostly in recruit and officer training institutions.
2. A parody of the 'fire and forget' missiles also called 'FAGOT'.
3. Pacing stick looks like a giant divider about three foot long and used by drill instructors to measure the length of a pace. Still affected by old time drill instructors more for show than actual utility.
4. 'Flatfoot' involves jumping till your knees touch your chest and crashing your feet back full on the ground with toes and heel making contact simultaneously. Bone jarring but effective for warming up in drill.

Chapter 17

TESTING TIMES

The one mile test had just concluded and most of the I termers were on the ground next to the finish line. Some were just catching their breath, most were hugging their shins. They had just met one of the new recruits' biggest nightmares, stress fracture, also called shin splint. This painful affliction was caused by repeated pounding of untrained feet on hard surfaces and the pain was similar to that of a fracture.

It was not a serious malady...depending on your point of view. It generally disappeared if the 'pounding on hard surfaces' was stopped. But that was not an option. If you reported sick and showed it to the doctor, he would get it X-rayed and give you a few days off PT and drill. If the pain persisted and the X-ray showed calcium deposits forming in rings around the bone, then the doctor may even recommend plaster. That would mean about a month off, and of course regressing to the original standards of physical fitness.

It could result in failing on the tests and being held back during term break for retests. The ultimate horror of horrors was being relegated. That was a fate worse than death and definitely not an option. No one could survive I term twice. There were very few cases of I term relegations...could be counted on the fingers of one hand. Most preferred leaving the IMA rather than going through I term again.

Most long-distance running was on roads. All drill was on the tarmac drill square, wearing hobnailed boots with steel-shod heels. With the drill instructors harrying everyone to slam their feet harder the situation only worsened. There was no way out of the hard pounding the tibia and the morale got. Most cadets just bore the pain till they could pass their tests. They then took it easy and of course, went home for term break to lick their wounds or have their mothers do it for them. Leave in the military was meant for physical and mental recovery. The fun was incidental.

Jay was one of those on the PT ground holding on to his legs waiting for the pain to go away. He was one of the lucky ones to have passed his one-mile mandatory I term PT test. The Academy took running tests more seriously than the others. The PT *ustads* were even known to reduce the pass time by a few minutes or seconds depending on the distance. This was usually done surreptitiously, with the encouragement and tacit support of the company DSs and IMA bosses.

The practice runs every Sunday under the supervision of an enthusiastic third termer had been hell. The III termer in question was keen on joining the Special Forces and becoming a para-commando. This cadet, GC CUO (Company Under Officer) Dhaiya, was physically gifted and mentally unstable...at least that's what most of his peers and juniors thought. He was physically very hardy, and it was almost confirmed mentally too.

GC Dhaiya tended to go on long runs at odd hours with a twenty kilo backpack just to toughen himself up further. The I termers stood no chance and had no way of keeping up with this Speedy Gonzales from hell. But Dhaiya was hell-bent on making all the I termers fit as commandos. The I termers had much lower aspirations. They just wanted to pass the lowly one mile test, "which any suitably motivated girl could run"—or so proclaimed cadet commando Dhaiya to the I termers. Kapoor murmured sotto voce, "One look at Dhaiya and any girl would be motivated to run fast and far!"

After a few weeks of running with an aspiring commando, the heavily or actually severely perspiring I termers could finish the one mile test. At least in satisfactory timing, this was about six minutes. Some could even do it in five, while Dhaiya could do it in four-and-a-half without any serious sweat.

There was however a sad soul, who for the life of him just could not run. This was GC Sandeep or 'Sandy' Ahlawat. He had been a BOSS in his previous life or 'brother of sexy sisters'. There were actually two sisters, both older and both quite sexy by ICS or Indian Cantonment Standards. He had been thoroughly repressed and mollycoddled in turns by his mother and two sisters. The end result had been a thoroughly unreliable sissy.

Sandy had joined the Army because his father, an Army doctor, had lied to him about the true nature of military service and Academy training. The fond father had imagined that the IMA would make a man out of a son whom he had treated like one of his daughters.

GC Sandy Ahlawat was described by his peers and superiors as 'handsome and tall...that's about all'. He did not like running and those running with him did not like the experience much either. He tended to get clingy and ask for help. He had put up a picture of his pretty sisters on his dressing table and that had elicited some offers of help initially. There were quite a few who tried to help him out in the hope of getting an introduction to his sisters. But they soon gave up. Even the sexiest and most willing women on earth were not worth the effort required to get Sandy moving...in a positive direction.

On the day of the final one-mile test, all the I termers took off the moment the whistle was blown, closely pursued by PT *ustads.* Cadet Sandy tried to sham it out by running at the tail end, hoping to take a shortcut. But wiser men than he had tried various tricks to beat the academy's system and the instructors knew almost all of them. As he had hoped, the main body of the I termers overtook him. They were closely followed by the PT instructors whipping the stragglers with their whistle slings. Gradually everyone overtook Sandy and he promptly dove into the bushes. His plan was simple and stupid. The bushes led to the path, which led to the culvert on the one mile route which was just 500 yards from the finish line. He would thus have to run just about 700 yards. The plan was fatally flawed since the last 500 yards would be packed with instructors encouraging the frontrunners to improve their timing. Also, there were only two sorts who took shortcuts in the Academy, in fact in life too; the very bold or the very stupid. In the military such creatures usually got gallantry awards. The former is quite rare though and nobody quite expected any I termer to be bold enough to try taking a shortcut. In a very short testing distance, by Military standards, of one mile which everyone closes in about six minutes, taking a shortcut is suicidal foolishness. But mummy Ahlawat had raised a handsome boy, not a particularly smart one.

As Sandy peeked out of the bushes, he saw a pair of red stocking-clad legs belonging to one of the PT *ustads.* Sandy withdrew into the bushes and lay as still as a shocked field mouse near him. Even the mouse had never seen or heard of a cadet taking a shortcut...in a one-mile test. The PT instructor waited till the last cadet (hopefully) had passed him and then left. Sandy was about to get out when he saw Capt. Shah ride up on his horse. There was nothing to do but lie doggo, even the mouse agreed.

Capt. Shah moved on and Sandy came out, cautiously. He was contemplating catching up with the rest of his mates when he was caught instead by a roving drill instructor. The conversation which followed illuminated the fact that even the dumbest human can get creative in the interests of self-preservation.

Drill Instructor in broken English[1] to cadet Sandy, "GC bushes in you what business?"

Cadet Sandy replies with a pained look, broken voice and equally broken English, "Fell I in bushes while one-mile test running!" He also clutched his knee for further effect.

Drill instructor gets even more suspicious and says, "Rest course GC where is... GC telling?" Apparently he was unaware of the PT test being conducted that morning.

Sandy grasped at this one straw, "*Ustaad* they running, running hardly... very fast like deer, I taking rest..." Sandy figured he would be better off speaking Pidgin English to put the Drill Instructor at ease. Yoda the Jedi master in the Star Wars movies would have been proud at this imitation of his English speaking style

The drill instructor however was not easily lulled. His suspicions were logical since even the dust the one-milers had raised had long since settled. Still there was not much he could do in the absence of hard proof. He looked hard at Sandy who gave him the most coquettish look he remembered from his sister's repertoire. His sisters had used that look often and gotten their way. He hoped it would work in this situation too.

The drill instructor was a bit confused. Here was this sissy-looking GC giving him an obviously bullshit story and weirdly inappropriate looks. Neither of which matched the given situation or time. One thing was certain as correctly identified by the drill instructor, this GC was lying and he was a shammer. Both unpardonable sins for a GC and this drill instructor would not let it go unpunished.

Just then a PT instructor came running up looking for GC Sandy. He had been missed in the head count after the test results had been compiled. The drill instructor and the PT instructor had a brief whispered conversation and a very meaningful look at Sandy's last resting place, the blessed bush.

Sandy knew he had partially won this battle, but it was going to be a very uncomfortable war. The next week's BRO[2] brought everyone some comic relief.

GC Sandy was awarded fourteen restrictions for 'Lying in the bushes with instructor'.

The story was not over, not by a long shot. In four-legged animals or maybe Arnold Schwarzenegger's family it would be fairly easy to find creatures with strong fore and rear limbs. The I termers, for the most part did not possess adequate upper body strength. They may have done a bit of walking or running, but very few had done anything that entailed upper body strength. They were not at all prepared for the chin-up test, as a pre-test sadly revealed.

The company commander was appalled and ordered the CUO and the CHM to do the needful. The company championship banner and his promotion to lieutenant colonel might be at stake. Thus the 'no questions asked' clause was unsaid and understood.

The I termers were in for a pleasant shock that night after roll call. They had some inkling because Bhasin, the BCA's pet, had overheard plans for some sort of night 'programme' for the I termers. He had, in a rare fit of course spirit, warned Jay not to eat heavily in the night. Jay duly passed on the warning and was duly ignored.

The roll call got over and then the CSM gave a short speech on the evils of being physically unfit. He particularly went into details on the misery that was likely to befall those who could not do at least eight chin-ups. JUO Akshay Kumar jumped up onto the veranda and said "Get on hands down, you fuckers!" And just like that evil befell them.

The I termers were vaguely sure they were being put through pain for their own good. They were only mildly concerned about the wet ground staining their suits and ties. These pushups would help them pass their chin-up test-the CSM had said so and they had no choice but to believe him. After the first 100 pushups or so however, Jay and the rest of the I termers did not believe the CSM so much. They were no longer convinced about the salutary effects of pushups on one's moral and physical well-being.

After a 150 pushups they didn't care about anything. The wet ground staining their clothes, they failing in their chin-up tests, never going home again, nothing mattered-as long as the unending pushups stopped. They were waiting for only one word 'Breakoff!'...or 'fuckoff'-both were welcome inspite of the singular difference in protocol.

After the 200 count, the pushups became mechanical. Even JUO Akshay Kumar gave the count automatically without any venom or vim. It was the same, the count, and you went down, then he would say up and you would come up, over and over again... Time seemed to have stopped and their muscles were numb and beyond pain. Unless they upped the ante and put a brick on your ass, the weight difference was just a kilo or so...but the effect on morale was a lot, lot more.

The third termers in charge of this little physical tuition stopped telling the I termers to do the pushups correctly. They had all been there... and right now the I termers were working on the outer boundaries of human endurance. The theory was if you made a man do hundreds of reps of any exercise, he would do at least 25 percent of them correctly.

After four hundred pushups, the CSM decided to call it a night. They had achieved their objective at the two hundred mark itself the next two hundred odd push-ups were just to toughen them up. Of course the only losers were the idiots who did it sincerely.

The I termers staggered back to their rooms. Kapoor nodded to Jay and they made their way to Ghai's room. Ghai opened up and nodded to them as they stepped in without saying a word. Ghai wordlessly opened his trunk and took out a bottle of brandy. They all took two swigs each and went back to their rooms and crashed, still in their wet and muddy clothes and slept the way their arms and shoulders felt, like the dead.

The night pushups continued for a week, and then came the day of the chin-up test. The I termers were actually relieved since passing the test would free them of the pain in the ass called the pushup. The pushup test had everyone passing but the chin-ups were a disaster.

The PT instructors were always very strict in the first attempt. They insisted on correct form and failed people ruthlessly. Officially speaking, attaining the 'satisfactory' standard was enough to pass. But the Academy expected prospective officers to be of 'excellent' standard, physically. After all, young officers had to be very fit. It was actually a good way to earn the respect of their men or soldiers, who were usually excellent in physical tests.

The PT instructors discounted any chin-up that was not correctly performed. Thus only four I termers from Bhagat battalion passed. Jay was not one of them neither was Kapoor or Ghai. Regrettably Bhasin and Sheoran passed.Being physically fit and popular were different issues.

They went back to their cabins dreading the next round of pushups. Jay fasted for one whole week, while practising chin-ups on his curtain rod. By the end of the week Jay could manage eight passable chin-ups. Keeping the body rigid would be crucial. The PT SM would be taking the retest and was a stickler for correct form.

Jay was in the line-up for the PT test and Sandy Ahlawat was in front of him. Jay groaned. Sandy was bound to irritate the PT instructor and thus ensure everyone following him had a tough time. Sandy did not disappoint. He kept simpering and giving ingratiating looks which only made the PT instructor and the PT SM suspicious and irritable. They moved in for the kill.

"Ok up!" said the PT instructor who happened to be the infamous Hav. K.K. Swamy aka '*kala kobra*', as he constantly reminded the GCs.

GC Sandy Ahlawat who was six feet tall and could easily reach the chin-up bar, hammed it out and pretended to jump for the bar. He also made annoying heavy breathing sounds intended to indicate strenuous efforts in reaching said bar. The PT SM lost his temper and asked the PT Havildar next to him "*Is* GC *ko* warm-up *nahi kiya*?" The PT instructor was offended and took the PT SM's comment personally. He promptly bawled "All test GC bend position!"

The hatred the test takers felt and directed towards Sandy was palpable. The GCs taking the retest were all hanging on by a thread (that was how the PT *ustaads* described their arms anyway). They had just enough strength and stamina to do eight 'demo type' chin-ups and live to retest somewhere else some other day. Every pushup they did now would subtract from their precious arm strength. The PT instructor was also aware of this and sympathised with them, but the PT SM was watching.

The number of PT failures reflected indirectly on the inspirational and instructional prowess of the instructor. Hav K.K. Swamy was keen on his personal career progression. But the *kala kobra* also had a compassionate side, despite the scary exterior and the general reputation. He genuinely wanted his cadets to get physically fit and pass the test. Otherwise they would be held back during term break, even get relegated and worse, bug him. He would be the one training them during the term break.

After about twenty measly poorly regulated pushups Hav K.K. Swamy bawled, "GC LINEUP"

There was actually no need to shout. These were the magic words all the cadets were waiting for. They jumped up and lined up fast. It was so fast one might have thought it was a nude supermodel rather than a chin-up bar in front of them. Their hearts could not have beaten faster if it had been a supermodel au naturale or a loaded barbeque stand...It's not always about meat. But every beat of their hearts thumped in rhythm to the words, 'Die Sandy die'.

The PT SM saw the sweat on the brow of the cadets and beamed his approval. His beam was slightly dulled when his gaze stopped at GC Sandy who had done only one or two pushups and still had his goofy smile intact. The old warhorse had seen many cadets and realised this one would take a lot of work, if he survived his coursemates' ire.

He pointed at Sandy and said, "You GC, fallout, test carry on!"

Hav. K.K. Swamy took up the slack and pointed at Jay who was now first in the line and said, "START!"

Jay grabbed the chin-up bar and waited for the count. He managed to do eight but the ninth he did the snake. This referred to the tendency to sway and jerk one's chin past the bar. It really did look snaky, the swaying and swinging up side to side. And the K.K. Swamy aka black cobra was not impressed. He just said, "FAIL!"

Jay was hustled to the side by another PT instructor who hissed in a not unfriendly tone, "Side jump carry on warm-up *rehna*, second chance..."

Jay's morale soared along with the adrenalin. He might get a second chance...he might get a second chance..he might get a second chance. The mantra ceased when the PT SM came over to Jay and the two-thirds who had 'refailed'...as the PT instructors put it. The SM just said, "Hmmm...ok, last chance!"

Jay and the other 'refailures' lined up again and this time Jay managed nine chin-ups. It took every last ounce of his will power and energy, the adrenalin helped too. Most of the cadets passed, the failures being motivated by the promise of a retest before the term break. The PT instructors appreciated sincerity and hard work; they knew the remaining failures would practise chin-ups till their arms stretched by an inch at least.

The expected exception was Sandy who was last seen trying to snake six chin-ups using mostly his backside, with the PT SM shouting his encouragement. He evidently still believed he could manipulate the laws of physics. Only Sandy would defy the laws of biology and tried to overcome muscle deficiency by using his ass to get his chin above the bar.

That boy had learnt all the wrong lessons from his sisters. Having a good ass could get you far in the world but it was doubtful if it could get your chin past the bar in a chin-up test. This cadet would clearly not try his best, but the PT SM and his staff would.

Jay rushed to his room without another look back. He had passed all his tests and now there was very little between him and his mom's cooking and...lots of sleep. Sandy and a few unfortunate others had, to paraphrase Robert Frost, 'miles to go before they could sleep'...literally miles...that too mostly running.

There was but one thing between him and II term-the final camp. As some general had probably or should have said—one battle at a time, one victory at a time...

NOTES

1. Drill instructors loved to use whatever little English they knew because they thought it made them sound like the officers. The end result was usually hilarious.
2. BROs (battalion routine orders) taken out routinely to update duties responsibilities, general info especially on punishments awarded among other things.

Chapter 18

THE VIRGIN CAMP OR SILVER CHAIN

Jay leaned wearily on his shovel and tried to make out where his trench mates were in the pitch darkness. He was supposed to have been digging for the first hour and then be relieved by his trench mates, Bansal and Sheoran. It was now nearly two hours that Jay had been scrabbling in the rocky ground without achieving much depth. He had lost all sense of time ever since he had given, or more accurately, Kapoor had taken his watch before going on guard duty.

Jay threw down his shovel and shivered as the cool breeze chilled him through his sweat-soaked dungarees. He wondered about his missing partners...masters Bansal and Sheoran. He cheered himself up by mentally rehearsing and reviewing the harsh abuse he planned to greet them with. That is if he saw them before exhaustion prostrated him. Talking about abuse, he was acutely worried about the harsh abuse the instructors would have to offer if they did not finish the trench before dawn. He was already exhausted; it was just the fear of the instructor that kept him digging. He missed the days when it was just fear of parental displeasure.

Bansal was a born shammer and cribber; he hadn't stopped whining since their final camp 'Silver Chain' had started two days ago. Sheoran had hurt his thumb during the inter-company boxing championship and was something of a prima donna. He had taken full advantage of the sympathy from the company officers for the I termer who had shown so much boxing skill. Never mind he had taken a pounding and had been defeated, Sheoran had managed to convince his platoon DS that he had lost because he'd hurt his hand. It didn't matter that none of Sheoran's course mates or seniors believed his theatrics. And now it was Jay who was directly affected by the Sheoran's 'single handed assistance'.

Jay thought back to their first camp. It had actually been just an overnight tent pitching competition. Then too, he had been unfortunate enough to be part of the group with Sheoran and Bansal. Those two had done 'sweet fuck all', to paraphrase a popular military comment.

With Bansal vanishing every so often for a break and Sheoran only offering valuable advice, it was a miracle the tent had come up at all. Jay remembered how they had slogged all night, smoothing out the wet mud in the freezing cold, while Bansal and Sheoran were nowhere to be found. They had been sleeping in the bushes and were chirpy and bushy tailed when the DSs came around. The instructors had come to check if the camp layout and alignment was as per regulation.

Jay and Kapoor had watched with repressed fury while Sheoran cornered the glory. The ass had actually given the visiting officers the impression that he had been the sole artist responsible for the magnificent Taj Mahal of a tent that had come up. He had narrowly escaped a '*kambal* parade' or 'blanket party'[1]. And now he was up to his old tricks again in the company of Bansal.

Jay heard someone coming; he got up, dusted his pants off and tried to make out if it was Bansal or Sheoran in the darkness. He picked up a stone and threw it towards the figure along with a graphic epithet that implied violent incest. The figure swore back as the stone hit him and he sped up his pace. It was one of the instructors. Jay jumped up and saluted, slipped and fell into the half-dug pit, scrambled up and stood to attention. He then bent down again to get his rifle. He was still frantically scrabbling around feverishly, when he felt a sharp burn across his shoulders. He looked up and realised the instructor had whipped him with his walking stick. The stick was marked and could also be used for checking if the trench had been dug to the correct depth.

"That's for the bloody stone you threw at me, idiot. I'm letting you off because you were alert, but next time you may like to challenge and ask for the password before chucking stones like a bloody *junglee*!" said the instructor with a smile. Jay was grateful he was being let off with just a light whip. That slight smile warmed him more than any roaring log fire. It could have been worse. The smile actually helped reduce the sting.

The instructor looked around the trench and measured the depth with his stick. He shook his head and said, "Doesn't look like you're going to get this ready in time for the 'stand to'[2]. Where's the rest of your team?"

Jay was sorely tempted to tell the instructor exactly where the two shammers were, but was bound by the code of 'course spirit' and so he said, "I think they've gone to dig the DTL (deep trench latrines, usually dug some distance away from the main post)."

The instructor stood and stared at him for what felt like eternity and asked, "Who are your trench mates? It's Sheoran and Bansal, isn't it? I guess that stone you threw at me was meant for them, huh?" Seemed the lazy duo's characters were well known and discussed in exalted circles. Yes...captains and majors were part of exalted circles when you were a cadet.

Said captain rubbed his leg and stood thinking for a moment. He seemed to make up his mind, then whacked Jay once more on the head with his stick and said, "That's one more for the bloody stone you threw at me that hit me on my damned shin...and also for not keeping control on your trench mates! How the fucks can you keep a control of your men if you can't control your coursemates! Carry on..." With that the officer walked off leaving a slightly pained, very relieved and somewhat slighted Jay. His leadership qualities had been denigrated, the biggest rebuke an officer-to-be could be given. Jay was justifiably miffed; after all the 'men' would be under command, but he had little moral and legal command over his coursemates. This the DS knew, but it was just his way of telling Jay to be more forceful with his peers. Sometimes one has to be more firm with friends than enemies.

Jay was happy to get away with just a stinging welt as he heard the instructor rousting out the next section that had been sleeping. Jay thought he had seen the last of him and got back to digging, hoping his trench mates showed up soon. They did, and not exactly in the way he had expected.

After an hour, he was taking a break when he thought he heard the instructor's voice. Jay jumped up and started digging again, keeping a wary eye out for the instructor. There were just two more hours left for dawn and he had a nasty feeling they were in for some pain for not having completed the defences in time. At least Bansal and Sheoran had had the opportunity to get some sleep. Then he saw the instructor approaching with two guys rolling in front of him. Jay felt the familiar mixed feelings that military men all over the world feel when seeing a comrade undergo punishment. They were feelings of pity for the one punished and of relief for not being the one punished.

Those feelings vanished when he realised that the front-rolling couple was

Sheoran and Bansal. It turned out that the instructor had actually gone looking for the duo and found them sleeping in the bushes. When questioned, they had blatantly lied about taking a break after finishing their trench. They had not realised the instructor had already finished his rounds. He had already seen the trench and thus caught them shamming. That and lying made it a double offence, considered extremely unofficerlike. The repercussions had been terrible for the duo, but left Jay feeling righteously requited. The erring slackers were feeling wrongfully abused and aggrieved and they piteously proclaimed the same.

The instructor left Sheoran and Bansal with some unwelcome suggestions, "Complete the trench at first light and yes, briefing on arc of fire and fire card[3] ...Get cracking, you lazy bums!"

The instructor had one last parting shot, "Oh, by the way, you will wear your packs without uniform or boots till the trenches are complete!" thus adding insult to injury, very neatly. Jay subdued a violent urge to giggle loudly.

On that cheery note, the three were left in the semi-darkness. Jay said, "I am going to get some sleep, the shovel and pickaxe are in the trench... get cracking like the instructor said..." With that Jay made a satisfying exit.

First light, d+2 or day two of the Silver Chain exercise saw a lot of frustration and anguish in the trenches of the I termers of the IMA's various companies, now reorganised into platoons. Strong feelings were rife in the amateurishly-made trenches of Bhagat[4] battalion. They were very similar to the fear and anticipation in the trenches at Flanders and Marne in the Great War[5].

In Jay's trench, the atmosphere was a cocktail of emotions like rage, terror and extreme anxiety. The three different emotions were felt separately and individually by the three gentlemen cadet inhabitants of said trench. Or maybe that excavation should be called by its correct name, 'pit'. Calling it a trench would be an insult to all those brave men down the ages who had fought, died or survived in one.

As soon as the instructor and Jay had removed themselves, Bansal and Sheoran had sat down. They decided to ignore the imprecations to dig and rather decided to take some time off and then some additional time off. This was spent licking their wounds and discuss the sheer injustice of a system that did not allow a human just a wee nap. As they spoke they dug, but soon the laws of nature along with their natural callousness came into play and they fell asleep. Their efforts had subtracted just a few inches of the additional two feet required. Twenty two inches additional inches that were vital for a trench that might come up to the Academy's satisfaction.

Jay woke up when he heard the buzz of activity that happened at dawn in any defensive post real or otherwise. He saw the grey streaks of dawn and scrambled for his equipment. This sort of action was standard in any Army camp. Jay shuffled up to the trench, stiff with cold and fatigue. But it was not the cold or his aching back that made him freeze; it was the sight of his two comrades sleeping in an extremely shallow hole in the ground. To call it even a foxhole would be extremely insulting to 'fox kind'. Jay jumped into the hole and frantically kicked the slumbering gentlemen.

"Motherfuckers! Stand to is in a few fucking minutes and the trench is not yet bloody well dug!" Jay said. He took a few deep breaths and continued as calmly as he could. "I'm going for a crap, fuck the trench and fuck you Charlies[6] ...we are so screwed!"

The 'Charlies' looked wounded and then had the grace to look sheepish. They realised they might have to face the instructor's wrath if they did not do something about the trench. They got to work and were ready by the time Jay got back after relieving himself physically if not mentally. He was not at all relieved by what he saw.

It was an old army theorem that if you want something done fast, give it to a lazy man. But the quality of work by a lazy man albeit executed fast, often left a lot, quite a lot to be desired. Sheoran and his sidekick proved that theorem by standing in full battle gear inside a trench that would have been right for kindergarten kids. They had piled stones around the edges, and were crouched on their knees, to ensure that only their head and shoulders were out of the trench.

Jay was speechless with frustration, but had no choice but to put on his field equipment, get into the trench and wait for his fate. It was not late in coming.

"All right who is going to brief us here?" asked the instructor, familiar from the night before.

Sheoran jumped out of the trench and confidently rattled out the fixed line arc of fire landmarks etc. Jay was impressed, the instructor was not. For one thing Sheoran was parroting a briefing, which was for another part of the defences. He had copied it from a sincere, hardworking and gullible coursemate the previous evening.

It was amazing how the Sheorans of the world always found suckers to serve every need. But there is a popular Army saying, 'you can fool some of the people some of the time, but not all the people all the time'. The instructor was among those not fooled. The DS had seen, met and worked with a lot of these 'smart Alecs'...he was posted there for that reason.

"You bloody fool at least have the sense to change the landmarks when you copy someone else's briefing...now get on your hands down!" With that the DS turned to Bansal.

"What is the correct width of a three man trench?" asked the DS

There was a brief pause before Bansal brightly chirped, "Three feet...". He realised his mistake as soon as he had opened his trap. The question had been rhetorical and thus the inevitable follow-up question came.

"So how many feet is this so called three man bloody trench..." asked the DS as he swung his measuring stick with a promise of menace.

Sheoran was quick to appreciate the gravity of the situation and volunteered with a martyred look,, "Sir, I've hurt my hand and couldn't dig much..." and trailed off, silenced by the contempt on the officer's face.

"Who made this range card?" asked the instructor

Sheoran answered instinctively, "I did, sir!" and realised he'd also made a mistake opening his trap, given the current atmosphere of impending doom.

"Seems you silly cunts haven't got the hang of judging distance and decided to copy the neighbouring trench's range card too..."

Sheoran tried gamely one last time, "Sir, it was for mutual support...overlap of arc of fire..."

"You've only hurt your hand isn't it not your bloody brain?" asked the instructor sarcastically. Sarcasm in the army is usually followed by straight forward insult and painful directions.

Two minutes later Sheoran was on the way to the nearest landmark for the fixed line, staggering under the weight of a double pack and three rifles. Bansal was rolling along the same path 'on the rocks' literally and figuratively. Well he was wearing his underwear. Jay was instructed to continue digging which he did with a relieved sincerity and diligence, which was very industrious by any standard. That marked the beginning of a very long and unfulfilling day, at least for Sheoran and Bansal.

Jay and his trench mates were not the only ones who were suffering the attentions of the visiting DSs. There were others who were treading the tear filled path of Sheoran and Bansal in various modes. Some victims were carrying weapons up, some rolling, others running singly, in pairs or even carrying another comrade. Much sweat was shed on that mock battlefield that day. Jay sighed to himself. The IMA or Army slogan 'The more you sweat in peace, the less you bleed in war' had better be right.

Sweat was something Jay was acutely aware of that day. He was soaked in it, it was trickling down his back in streams; it was stinging the cuts and bruises on his back because of the packs and the rolling. It was dehydrating him during day, and chilling him at night. He hated sweat. The night had been chilly and the day was beginning to be hot and sultry. The others were suffering bruises and cuts rolling on the stones, and all were in various states of exhaustion. Jay had been amongst the fortunate few to be 'allowed' to dig the trenches as part of the penance. After one hour of digging, he wondered if it was so fortunate after all.

The instructors were everywhere, screaming oxymoronically, 'C'mon you sissy buggers! How are you going to fight a war if you are going to get shagged out so soon!" or words to that effect. The whole action seemed pre-planned. The single common thought going through every cadet's mind was, 'Shoot me now and let's get it over with' or thoughts to that effect.

Discomfort had diverted the mind from other human needs but after the instructors had departed, other basic instincts surfaced. A couple of fist fights broke out, and were broken up by other coursemates. Everyone was hungry, whatever breakfast that had been dumped at the defences was cold and had sand in it. It had been left there by the kitchen lorry while they were being entertained by the DSs.

The subdued I termers ate a few mouthfuls and gathered up their energy for the next round of extreme physical exertion.

The final piece of the 'Silver Chain' exercise was the 'run back' or route march or forced march as it was variously called. The distance to be covered was only 25 kilometres. This distance was not much to a trained athlete and definitely not for a trained and bloodthirsty soldier. But for new recruits it seemed like a very long journey, even on wheels. They had enough time to mull over that.

The senior instructors came around to see the effects of their morning ministrations amidst renewed nervous anticipation. But they just looked around and left after giving instructions that the trenches should be filled up and all traces of cadet inhabitation removed. The time available for that was one hour after which they were to gather at the adm (administrative pronounced as adam) area where the kitchen lorry would meet them for the final meal.

The pack of I termers wearily dragged their packs and digging tools back to the adm area and fell in to give a report to the duty JCO. It was 1100 hours. It turned out that a rifle magazine was missing. They all went back to look for it. It was found at 1230 hours. A colonel from the IMA training team, who was passing by, observed that the trenches had not been satisfactorily closed. His observations were duly conveyed, rather forcefully to the cadets. And they went back to work on closing the trenches.

The defences were finally smoothed out to everyone's satisfaction...well not everybody. Bansal and Sheoran had done the job of filling up the trench with the same sincerity as they had in digging it. The dominant wish of the group was to close the trenches with Sheoran and Bansal in it. They got back to the adm area at 1330 hours and lay down in the mud to wait for lunch. It was better not to think beyond lunch if one wanted to have an appetite. The thought of the route march gave pangs of anxiety rather than pangs of hunger. Covering 25 kilometres was expensive even in a taxi or auto...

NOTES

1. Blanket or *kambal* parade/party is an old organizational method internationally recognized, to deal with non-conformists, slackers or malingerers who refuse to abide by accepted conventions and /or bring disrepute to the organization. It can vary in its intensity and violence and is usually meted out by peers with the blanket covering the person, thus signifying the wrath of the entire group without any single individual being identified. It is officially banned but still carries on and is at times, highly effective.
2. Stand to-refers to the manning of weapons and trenches, usually done when a threat is imminent and usually before dawn or dusk when attacks are most likely or also when some VIP is visiting.
3. Range card is just a simple card that has various landmarks on your arc of responsibility marked with compass bearing and range to make it easier to engage targets if attacked. Simple to make for a trained soldier but cadets and recruits find it mind-boggling.
4. The battalions were named after ex-Chiefs of the Army, while the companies were after battle honours. Bhagat battalion was thus named after Gen. Bhagat.
5. The World War I is referred to as the Great War and casualties of the trench warfare at Marne and Flanders were particularly unprecedented.
6. Charlie is a derogatory term synonymous with 'loser'.

Chapter 19

LIGHT AT THE END

There was much discussion and thought given to how a route march could be completed in minimum time and discomfort. The word ambulance was mentioned with inappropriate frequency. Someone declared that the instructors made the whole group repeat the march if anyone was left behind. No one knew how true that statement was but it matched the IMA code they had experienced so far.

Sheoran, who had cousins in the Army, started a loud monologue on the techniques of completing route marches comfortably. Those within earshot perked up. When hope is low all advice sounds good. Jay shook himself awake to listen to Sheoran's pearls. The pearls were mostly nuggets of pure bullshit, but everyone still listened. They listened mostly because they were too raw to recognize the smell of bullshit as yet. They also listened because they were in the 'Nothing to lose' zone mentally.

All those present had heard many tales of horror. They had seen their seniors come back with blood-soaked socks, blistered feet, bruised backs with pack strap burns on the shoulders. The I termers of Bhagat battalion were very apprehensive. Their platoon DS had helpfully informed them that they had to finish the route march in five hours. A quick math brought a chill up their spines, when they realised that meant running at the rate of 12 minutes a kilometre with a full 15 kilo pack.

Twenty five kilometres could be completed in three hours by any reasonably fit platoon. The distance did not mean a whole lot to any professional soldier, but for the semi-trained gentlemen cadets in their I term, the hard truth of soldiering was finally coming home. Without taxis or credit cards emulating

Rambo or Arnold in 'Commando' meant a lot of sweat, more running than shooting and definitely no heroines anywhere. It was going to be a long road.

Among Sheoran's pearls of wisdom was a gem. "When running long distances, one should eat light and make sure the bowels are clear!"

Jay digested that information slowly, but not clearly. There was actually not much else for him to digest. Very little of last night's dinner had reached Jay. It had been dumped at the adm area and mauled by a nearby platoon. Scavenging cadets and stray dogs had taken whatever was remotely palatable, what was left had too much sand in it making it unpalatable for even the most easy-going hyena.

Breakfast had been skipped due to the exigencies of service and the exacting defence standards of the Indian Army. Lunch arrived just as Jay was considering the merits of a light stomach versus a full one and bliss. He remembered the guys puking on the drill square or firing ranges after heavy breakfasts and consequently heavy exertions. He decided to skip lunch except for the dessert which was some melted ice cream and could be drunk like a smoothie. And that was describing it positively...

The only worthwhile piece of advice they got was from the store NCO who came to collect the tools. The best way to avoid blisters during a long route march was to wear two pairs of socks. A skin tight nylon sock inside and cotton or wool over that; the theory being the nylon socks would prevent the bunching up of the socks and act like a second skin. Unfortunately most of the first termers had not thought to bring a second pair of nylon socks. They second pair they had brought was the woollen issue one, as authorised in the manual.

The run was about to start. The Bhagat battalion I termers stood around, adjusting their packs and fidgeting nervously. They were the last to be released. They would have preferred a different kind of release, maybe from the Army itself. Many looked longingly at the ambulance. The instructors watched the anxious I termers with detached amusement. This was just a 25 kilometre hike, no big deal.

The I termers of Bhagat battalion started off with a big rousing yell and a fast pace. Actually the yell was a bit shaky and not as rousing as it should be. The bored DSs were however roused. They made them come back and give a few more cheers before they were satisfied with a decibel count which indicated enthusiasm. The pace started tapering off after a kilometre.

The first cadet to give up and start walking was Bansal. Sheoran lagged behind and then pretended to be helping Bansal along when he saw a jeep approaching. He hoped there would be a DS in it who would be suitably impressed. It was a DS alright but not suitably impressed by Sheoran or Bansal and threatened to kick them into a higher gear or speed as required. It improved their speed but just till the DS was in sight. By this time several others had started to shuffle a few steps of mock run and then walk. This slowly became more of walking and then deteriorated into shuffling.

Jay was one of those who could not keep running. He was feeling nauseated and just could not summon the energy to put one foot in front of the other. The lack of food and exertions of the last few days was hitting him and some of the others in the form of extreme exhaustion. It was a new physical state for almost all of them. The others saw it as shamming and tried to drag the stumbling cadets along. The DS kept driving by in their jeeps and exhorted the heroes to pull the weaklings along or the entire group would be forced to repeat the route march. The threat worked briefly.

'Helping' the weaklings or dragging them along continued for another 10 kilometres before almost everyone gave up. Jay somehow managed to keep moving. He stopped to pee and was horrified to find he was urinating dark red liquid that looked like blood. There were only eight more kilometres to go. Jay could not bear the thought of going through the agony of I term or the camp again. He had to pass this term and he had to complete the route march for it...no matter what. He shambled across the finish line in a semiconscious state barely aware of the red stain on his pants crotch.

He did not have the strength to report sick. He went back to the platoon in a daze, had the snacks and juice the seniors traditionally organised as welcome. The he just staggered to his cabin and fell on the floor and slept for 12 solid hours.

"Come out fucker!" This peremptory phrase was accompanied by persistent pounding on the door and the insistent phrase, "*jaldi, jaldi aaja bhosadike*!"

Jay dragged himself to the door and opened it. It was JUO Akshay Kumar. Jay's heart skipped a beat and his beaten spirits sank even lower. Maybe the pain-loving JUO had been sent by the DSs to sort out the I termers for not doing well on the runback.

"What's up fucker... busy shagging huh?" JUO Akshay Kumar at his uncharacteristic and diplomatic best. At least he said it with a smile. Jay had seen that smile frequently and it had frequently meant the worst. This was probably the first time he had seen him smile while vertical, usually it was while horizontal, doing pushups or something worse.

He saw the other I termers of his platoon, Bhasin, Shiju Abraham and Arjun Shinde also grinning widely. The smile on JUO Akshay Kumar's or Bhasin's face did not convince him, but those of his other coursemates did.

"Come on you bugger we gotta go for the treat!" said the now genial JUO.

Jay hesitated and then remembered it was a tradition for the senior terms to organize a party or a treat for the cadets back from a camp. This was done after getting back from the final exercise or camp route march of the term. This was the beginning of the end and there really was some light at the end of the tunnel, thought Jay. He put on his games dress and limped out. They all limped along behind the JUO like a pack of wounded ducklings.

The lunch was not really a success even though most of the I termers fell asleep while waiting for the waiter and then again during icecream. JUO Akshay Kumar, though comporting himself genially, was at times unable to fully prevent his true nature from peeping out every now and then. He slapped Jay awake, albeit with a smile when the food arrived. He poked Shiju awake with a fork and generally tossed out threats of pack parades as punch lines. He thought he was being humorous but it had just the opposite effect. It felt somewhat similar to a rabid dog wagging its tail and acting friendly all of a sudden.

Kapoor in his exuberance on seeing the glimmer at the end of the tunnel made the mistake of telling JUO Akshay Kumar a joke. The worthy Akshay Kumar smiled in his usual wolfish way. It took away appetites and further visions of freedom. They knew Akshay Kumar and knew he had something up his sleeve. The rest of the meal was in subdued silence. Akshay Kumar was the one who enjoyed the meal the most.

There was a tense moment when the bill came. They could see the consternation on their host, JUO Akshay's face. He was clearly overawed by the bill and his small mind was regretting the generosity of tradition. Tradition that decreed III termers always paid when dining with I termers especially at the end of the term. Still he gritted his teeth and paid up; after all tradition could not be ignored.

The bill would be split by all the third termers, and that was mildly cheering. He was the original cheapskate and the I termers noted this with disgust. One can tolerate a tyrant as a boss as long as he shows he has a big heart off-duty.

But JUO Akshay Kumar would never endear himself to his subordinates because he remained petty minded on and off duty.

JUO Akshay had one last laugh up his sleeve. "You guys better gear up for the inspection!"

Seeing the unspoken question on his audience's faces JUO Akshay Kumar elaborated, "The Academy is being inspected by the Central Army commander next month and the commandant is pissed. He has ordered cabin cupboard inspections to continue till then. " He then looked around triumphantly to see if his audience was impressed with this piece of inside knowledge.

Shiju, Shinde and Jay felt the food they had eaten turn into lead and the taste in their mouth as metallic as blood.

There were almost two weeks left for the term to end. Those days were traditionally filled with informal get-togethers and relaxed times for all terms. Except for those who had to retake tests or finish up punishments. But continuous 'cabin cupboard' meant no freedom and instructors hanging around all the time. This was bad news. The I term was still a long way from over. Akshay Kumar was the happiest man in the group. They did not call him 'Keen' Kumar without reason.

Jay went back to his room and started thinking. It was more than 12 hours since they got back from the camp. He still found peeing a painful experience and there were still tinges of blood in his urine. He decided to ask the 'Lancee' aka Lance Corporal Rajesh Pandey.

The Lancee's cabin door was unlocked. Jay knocked and waited, there was no response. Jay pushed open the door cautiously and walked in. Lancee Pandey was sprawled on his back with a cigarette. It smelled like some hallucinatory weed had been cremated with human remains in the room. Jay waited for a few minutes for Pandey to respond to his greeting.

"What the fuck do you want?" said the Lancee after a full two-minute silence.

Jay wondered if this was the right time and person to seek advice from, but he decided to bear on. "Sir, I've been peeing blood since the camp...should I report sick?" That got Pandey's attention.

Lancee Pandey sat up with a jerk and shook his head. "Fuck! How long has this been happening? Did it start after front rolling in some fall-in!"

Lancee Pandey's tiny, well-meaning mind with weed lubricated wheels was spinning faster than it was designed for. Evidently the grass had turbocharged his thinking. Fear of involvement in a potential 'ill treatment of subordinate' case however sent his thinking machine hurtling down the wrong track.

Jay described the origin of the situation and his primary concerns. Namely, if this could be a life-threatening situation or worse, something that could get him relegated or kicked out of the Academy. Also, last but not least, would it affect his reproductive capabilities? These were all genuine concerns to a young man still a virgin and a minor. Jay was but 20 years old.

Pandey thought long and hard and finally the penny dropped and he made the connection between the stress of the route march and the 'blood in urine'. "It's a common thing after a route march or sometimes even after cross-country or the punishment run[1] ...it'll clear up on its own after a few days..."

"But on the other hand consider it God's gift..." continued Lancee Pandey as he lit a fresh fag. Jay was mystified how blood from his most prized organ could be considered 'God's gift'.

"Listen fucker, go get yourself admitted and relax for a few days at the MH...unless you're keen on doing 'cabin cupboard' and tolerate inspections till term break!?" said Lancee Pandeyji through a cloud of cigarette smoke.

Jay looked at Lancee Pandey with suitable awe and gratitude for having shown him the true path. That was the true job of an immediate senior in the army. Jay went straight away to the JUO and told him he wanted go on special sick report. The JUO agreed readily when Jay told him his reasons to report sick. And so it happened Jay found himself in the section hospital, looking across the table at the sleepy young RMO (regimental medical officer).

"You bugger....you sure you want to get admitted?" The doctor asked with a yawn. Jay nodded hopefully. He hoped he hadn't looked too eager. But the doctor was too tired physically and mentally of military bullshit to care. He had had a hard day with his boss who thought he was soldier first and a doctor later... way later. He signed the admission forms and said, "You can get admitted tomorrow...now fuck off before I change my mind!" Jay jumped and did as he was bid.

"Sir one last question...?" asked Jay at the doorway.

"Yeah what is it?" The doctor asked, looking up.

"What's the reason for this blood in urine thing... is it curable?"

The doctor looked at Jay suspiciously and toyed with the idea of scaring him with some STD lecture, then took pity and said, 'It's because due to stress of extreme exertion, the blood cells break down and get flushed through urine...nothing serious, don't worry."

Apart from a burning sensation in his nether regions, Jay slept peacefully. As a coursemate seething with resentment at Jay's good fortune and impending good time commented, "It doesn't matter what's burning, as long as it's not your ass..."

That night Jay slept the sleep of the dead, without dreams, except towards the end when he had a short dream that he was in a tunnel rushing towards the light which turned into an ambulance and later a sexy blonde nurse with a Malayalee accent. The blonde part woke Jay up since it was so blatantly unreal. Reality was a fat motherly nurse but that came later.

NOTE

1. Punishment runs were part of the regular and legal punishments which involved a five-km run with full packs.

Chapter 20

Tender Mercies

Jay reached the IMA section hospital and was a bit surprised to see the 'sick report' crowd, many of whom looked suspiciously healthy. Bad news did travel fast and word of the impending pestilence of 'cabin cupboards' had indeed spread. The instructors, anticipating the cadets' attempts to flee bad times had also come there. They came to dissuade wannabe patients, torment a few obvious malingerers and generally amuse themselves. There is a bit of sadism embedded in every military soul.

Jay's heart skipped a beat; after all he was here to get admitted. All he had to do was show his hospital admission form...and then his heart beat faster. He saw Capt. Jeevan Singh leaning casually against the SH balustrade, looking slightly amused. That meant he had already had some amusement and was looking for more. A few more DSs joined him, eagerly scanning the horizon for more victims. Two battalions had already marched in their sick report and some of them were in pushup position now wishing that they were really unwell.

The military was probably the only profession in the world where sometimes not being fully healthy could be a relief. This revelation came to Jay as he marched into the hospital compound feeling like a gladiator. Capt. Jeevan actually licked his lips as he saw them coming. The entire sick report felt a cold sweat bead on their foreheads, mouths go dry quickly and other sub-conscious manifestations of fear. Some could not even manage this and just licked their lips in nervousness.

"Ok, what's your problem?" Jeevan asked mildly of the first GC who happened to be a III termer. The junior terms were not considered good sport.

The GC's reply was not satisfactory enough. He said something vague like stomach ache. In Capt. Jeevan's expert medical opinion, the GC's illness did not prevent him from doing pushups. Other members of the sick report began to feel, well, sick.

Capt. Jeevan called on the drill instructors who were hanging around, eager for a slice of the action, and said, "Check *karo in sahiban ka kya* problem *hai*?!" The II termer next to Jay perked up as he saw the Drill NCO come towards them. The drill instructor was a Gurkha and known to be particularly inflexible. So Jay couldn't really understand his neighbour's cause for cheer. The II termer was from Cassino company as his collar tab indicated, with the name tab, Vikram Sanyal. He saw Jay's incredulous look, winked and said by way of enlightenment, "Just watch, I tried the same thing at NDA..."

"What problem?" asked Hav. Tej Raj Gurung of the 1st battalion of the Fifth Gurkha Regiment nicknamed as Pfiffers, 150 years ago. Hav. Tej Raj Gurung was however nicknamed 'Viper' because of his fast snappy movements and deadly bite. It was not very clear if it was 'viper' or 'wiper' as in wind screen wiper. He did tend to weave a lot while the feet remained rooted at the same spot.

"Sahib..." said GC Sanyal ingratiatingly using a form of address meant for JCOs, "I have photosynthesis in my toes..."

Being addressed by the form reserved for JCOs, by GC Sanyal, did not soften the drill NCO. Hav. Gurung, short, wiry and vicious as a viper and about as intelligent, took a moment to ponder over this new medical condition, photo-something. He covered his momentary confusion by directing his attention at a few patients who were already getting his brand of medicine.

"Pushups continue... shammer GC!" said Hav. Gurung in pidgin English, to one of his other victims, while he tried to figure out a suitable response.

"You shammer GC!" this was addressed to GC Sanyal who stiffened up without losing his ingratiating smile.

"Photosynthesis not foot, in hand *mein hota hai*," tried Hav. Gurung bravely.

By this time the entire sick report, even the ones suffering punishment on top of other sufferings, was evincing a keen interest in the quasi bio-medical discussion between the 'viper' and his prey.

The exchange was getting interesting with Sanyal insisting that it was lack of chlorophyll that caused photosynthesis in the feet.

The strain of decision-making could be seen on the 'viper's' face. He was about to order GC Sanyal on his 'hands down' when he noticed that the entire lot of GCs and instructors were looking at him. The 'viper' came to a decision and he waved Sanyal on, into the infirmary. Some cadets at the back clapped discreetly at this minor victory over oppression.

It wasn't a wise move as the 'viper' lost his cool and rushed around to the rear file looking for the clappers. The clappers had been joined by the gigglers and the viper was foxed. He had but two fangs, pushups and the power to award extra drill parades. His venom could only be used against a few victims, not an entire herd. The scene was getting positively mutinous and Hav. Gurung was turning redder than his duty NCO sash, when help appeared from an unexpected quarter.

The senior doctor of the IMA section hospital, a crusty old colonel named Mehdiratta, came out and bellowed, "What the hell is holding up the bloody line?"

The other instructors perked up. Normally the officers would not have cared a hoot for a burnt-out GDMO (General Duty Medical Officer). But this one was the IMA commandant's old unit Regimental Medical Officer and now bridge partner.

It was Jay's lucky day, indeed for every cadet on the sick report that day. The DSs, deprived of their sport, stalked off after telling the drill NCOs to break off. There were smiles on every face which froze as the 'viper' turned around and looked at the Bhagat battalion group memorising the faces for future retribution and then he too slithered away on his cycle.

There was further good news; Col. Mehdiratta had a matinee bridge date and was eager to get through the sick report. He handed out generous doses of SIQ (sick in quarters), or Ex PPG (excused PT, parade & games). Those he was not sure about, he sent to the Medical Hospital, MH, for admission.

GC Sanyal on entering the doctor's office had wisely changed his complaint from photosynthesis and lack of chlorophyll to weight loss and lack of appetite. He was also on the MH bus along with Jay. There were five other suffering GCs, but the mood on the bus was upbeat even for those too sick to sit up. The sick report paused long enough to see the hospital types exit the IMA gates. If envious looks had been missiles, the MH bus would have been blasted to a crisp.

The 'hospital types' reached the MH, which was an old building located on a scenic hill. They were issued pyjamas and shown to their beds by a somewhat cute nurse and a grumpy corporal from the AMC (Army Medical Corps). The latter were also known as NAs or nursing assistants and usually carried a massive chip on their shoulders. This was characteristically manifested through grumpiness. It was rumoured that the abbreviation NA stood for 'nurse's arsehole' and hence the associated demeanour.

The term 'somewhat cute' was used with very little justification to describe the nurse's charms. The GCs had not been in contact with the fairer sex for some months now and anything vaguely female looked cute to them. The nurse was a short, squat, young South Indian, but she was aware of the Ibid phenomenon and smiled coquettishly at the GCs. GC Sanyal promptly claimed her for his own. He was the senior-most and probably the healthiest, so there was not much protest.

The GCs settled down and within a few days, many recovered from their ailments, real and imaginary. The doctors were a tolerant lot, having joined the Army after a longish stint in civilian medical colleges. Some were from AFMC, the Armed Forces Medical College, which was famed for its relaxed attitude towards its cadets. Indeed it was run somewhat like a medical college and rumour had it that there had even been a strike. In sum, most doctors, even the military surgeons, who were traditionally butchers anyway, did not really approve of the remoulding process at the IMA. They did not mind a cadet shamming it out for a few days. But just a few days...not longer.

Jay had recovered from his 'blood in the urine' and the only 'burning' sensation below his belt was caused by his lurid fantasies and the young nurses on duty.

It was just the extreme stress of the moment that had caused the red blood cells to break down and get flushed out. After three days of enforced rest and intensive testing, all the tests had come back NAD or 'no abnormality detected'. Jay was apprehensive; the axe was about to fall any time now and he would be back in the academy scrubbing his own, and maybe even other floors, for the inspection.

He went looking for GC Sanyal and found him placidly munching an apple and reading pornography under a tree. Life had been good to Sanyal. He was naturally skinny and thus had no trouble persuading the doctor that he was losing weight abnormally. The doctor's plan was to feed him a special diet for two weeks before reviewing the situation. Two weeks of extra rich food and no work in idyllic settings was a dream unattainable by most of humanity. GC Sanyal had achieved that at the tender age of 20.

Sanyal may have been of tender years but he had been in uniform since the age of 12. He had been in Sainik School, then NDA and now IMA. He knew how to survive in a concentration camp type milieu. He was widely known as the man with the plan and thus was found sitting with a troubled Jay.

Sanyal heard Jay out patiently, then thought carefully for a minute and said "No problem...but it will cost you." He added, "Just one new *pondy*[1] ..." after seeing Jay's face fall.

"See here's the situation," Sanyal continued, "you don't have blood in your urine anymore and there is no sign of kidney damage, cancer or STD right?!"

Jay shook his head sadly, "Yes absolutely healthy. You think they may run some more tests?"

Sanyal shook his head, "No way old boy, your condition is very common in training centres and the doctors know that. Your time is up unless..."

"Unless what? What!" asked Jay, hopefully.

"Unless you get some traces of blood in your urine at the final check," said Sanyal.

Jay was awed by Sherlock Sanyal's insight into the system's predictability, but had one more niggling doubt that needed to be cleared.

"Sir, how do we get the blood in the urine?"

Sanyal sighed, "Let me know when you're getting married, you will probably need my help there also, you dumb prick!" He pulled out a pin from his shirt and handed it to Jay, "Use that to prick yourself...and I don't mean that kind of prick this time...put a drop of your blood in the sample! Now fuck off and get me my pondy!"

Jay rushed off gratefully to manage some pornography for the great Sanyal. He thought long and hard on where to get it from and the answer came to him. He went to the Officers' ward and was welcomed by the three officers admitted there. Their ailments were varied and somewhat typical—jaundice, motorcycle accident and football injury. Their reaction was also typical of most animals, towards their young ones.

Jay was offered a seat, his health enquired into and Jay felt part of the Army family. This emboldened him to request for a *pondy*. The motorcycle accident guy gave him one. He had after all broken his right arm and was thus ill equipped to fully appreciate it. Jay got up to leave but there was one more task he had to carry out to seal the deal.

"*Ek kaam kar yaar,* do me a favour," said motorcycle victim with a snigger, "Tell that nurse I'm in need of a sponge bath, will you!"

Jay said, "Yessir" brightly, but he had his misgivings. He had a feeling he was being misused but he was still raw enough to not differentiate between a request and an order. He thus repaired to the nurse's station and found the matron, the senior-most nurse, also present there briefing the young Lt. Rosa Kutty, a new nurse. The matron was a crusty old battle axe named Sujatha wearing a Lt. Col's rank. She had seen her share of frisky military creatures of all ranks, shapes and sizes.

Matron Sujatha grunted, "What?"

Jay had the overwhelming urge to get discharged. He was also extremely aware of the pornography in his hand. He desperately rolled it up and hoped Madam Sujatha would not ask for it. She just looked at it intently till it grew warm in Jay's sweaty clutches. She probably knew what the book was about; after all young military men are not given to serious intellectual reading.

The matron broke the silence with another, "What?" This time the tone was gentler than it might have been with a young Lt. or Capt. They were all after the same thing, only differing in malignance.

Jay stammered and said, "Ma'am, they are calling for the nurse...in the...officer's ward..."

The matron sighed, Rosa Kutty looked down and Jay blushed.

"Yeah, yeah....Ok who wants a sponge bath now?" said the matron who recognized a familiar situation and then continued, "Oh never mind, I'll go and see myself!"

Nurse Rosa Kutty looked relieved, Jay looked horrified. The matron read their expressions and said, "What? They want a sponge bath and they're going to get one...I'm a nurse too, you know!"

Jay realised the matron must have sponged or bathed many a fevered brow and dampened many an eager libido in her time. The Officers' ward would be in good, even if somewhat callused hands. He escaped as gracefully as he could after collecting his lab testing form.

The next day he dutifully went to the lab and gave his urine sample along with the furtive drop of blood. It was not easy with a test tube in one hand and a pin between his teeth. He had never realised skin could be so tough.

Jay managed to get the hang of mixing the blood and urine cocktail for the pathologists over the next few days. His trysts with the lab became a daily affair as the doctors grew steadily worried about Jay's case. Jay was however blithe and content with the life the MH had to offer. By this time, GC Sanyal had organized a regular supply of booze and chicken, for a profit, of course. He had managed to recruit a hospital sweeper through vague promises of a promotion and other equally undeliverable but definitely ambiguous assurances. But who cared; every evening was a party because none of the nurses wanted to or duty doctors bothered to come and check the cadet ward.

Monday was the day when everyone would care. It was the MH Commandant's round, and every case would be reviewed. Jay was hopeful that his test results would earn him at least another week at the MH. The fateful day came and all the patients were ordered to sit on their beds like good children, especially in the Officer's ward which had more than its fair share of bad boys.

The ward was one of the most difficult to inspect. The officers were prone to jokes, irreverent comments and outbursts of giggles. None of which was in keeping with the solemnity of the occasion or the sobriety of a hospital. This was a military hospital, where people are supposed to be sick and miserable.

The Commandant's weekly round was one of the few 'official' moments in a military hospital with its semi-military type personnel. The AMC guys tried to take it seriously while the rest of the army did not share their enthusiasm and took it with a pinch of salt and a lot of humour. The Commandant usually gave the Officers' ward a perfunctory glance and moved on to other wards. He had been informed that this particular batch in the ward was quite... rambunctious.

But the cadets were not feeling their best this Monday. They were all suffering, physically and mentally. Both manifestations of symptoms were because of a serious hangover, caused by over-imbibing of the cheap yet overpriced booze 'managed' by Sanyal. Jay had swallowed a couple of mints to cover up the booze breath, but he was worried about his bloodshot eyes. He was not reassured by Sanyal's sage pronouncement, "Bloodshot eyes are a sure shot indicator of serious illness!"

The Commandant was a kindly, well-meaning old doctor. He was an eye specialist and still a bit unsure about how to deal with his recent elevation to the General cadre and its administrative responsibilities after becoming a brigadier. He came over to Jay's bed and asked him in an avuncular fashion how he was feeling.

Jay instinctively said, "Fine, sir!"

The Commandant did not seem very happy with the reply. He had seen and heard seriously wounded soldiers say they were fine. He turned to the gaggle of doctors around him and asked, "What is the problem with this young man?"

There was some hubbub as Jay desperately tried to overhear what was being discussed about him. Sanyal was winking at him madly trying to cheer him up. He did manage to hear the Commandant raise his voice once to say, "Two weeks? Two weeks blood in the urine..."

And then the crowd parted and the Commandant patted Jay on the back and said the dreaded words "Don't worry son, we'll have you out of here in no time!"

Jay's heart sank when he heard the Commandant ordered his doctors, "Give this young chap ten bottles of glucose and a diuretic, let's flush his system out and then see!"

What made it worse was that he said it with a genuine smile. The other doctors also smiled but it was the smile of evil humour to come. It was around this time that Jay's balls also dropped and joined his sagging spirits.

The next day dawned bright and beautiful but not as much for Jay. It would ultimately be a most testing day for Lt. Rosa Kutty of the Military Nursing Service. At 0900 hours sharp, she and her sidekick, the grumpy nursing assistant Nk. A. Pandey reached Jay's bed with the diuretic injection, glucose and stand. The other denizens of the ward were looking on eagerly. Sanyal had occupied the front seat and was giving encouraging winks to Jay and casting surreptitious leers at Rosa Kutty.

"GC Jay, you will feel like urinating very frequently so you can... um, keep your pyjamas open. If you have any problems, please ring for assistance and Nk. A. Pandey will be there to help you," said Rosa, all in one breath. She was in an understandable hurry to get away from the awkward situation.

And before Jay could react, Rosa Kutty had, in swift economical movements, injected Jay with the diuretic and hooked him up with the first bottle of glucose.

"Ring the bell when the bottle is done," and with that she was off.

Nk. A. Pandey looked sullenly at Jay for a minute, then slammed the bed pan next to him and said, "Nurse *ko hi* call *kar lena. Main* office *ja raha hoon.*" Rosa Kutty had rightly called him Nk. A. Pandey thought Jay. The grumpy NA had already been rechristened 'Ass Pandey' in the place of A. Pandey by the other GCs. The thought cheered him up till the first bottle got over. Jay rang the bell and Rosa Kutty promptly came over to change the glucose bottle and adjust the drip.

The nightmare took a turn for the worse when, while changing the glucose bottle, Rosa Kutty tried to help Jay untangle his pyjama strings. Her hand accidentally brushed against Jay's overworked organ, and he immediately had an involuntary hard-on. Jay was in a social and military quandary. He was extremely embarrassed and also fairly certain that he had committed some offence punishable by the IMA honour code. That could be dealt with later but the hard-on was here and now. He just hoped that the heroine of this tragicomedy, Rosa Kutty had not noticed.

Actually she had and was confused whether to be amused, honoured or insulted. She chose all three and did not show herself for the rest of the day. It was a nightmare for Jay to get the bed pan and untie his pyjamas all with one hand. Jay spent the next two hours traumatised, rushing to the bathroom with the stand attached to one hand, holding up his pyjamas in the other and then peeing.

Jay's comrades were highly amused and even took bets on how many times he would pee before the bottle had to be changed. By the end of the whole exercise, Jay had lost all and any weight gained at the MH. He was ready to go back to the Academy and to honest living.

But the chapter does not end here.

The next day an exhausted and washed out (literally) Jay came back from the lab after getting his urine checked again, this time without any bloody contribution. He was to be kept under 'observation' for the next three days before being discharged. When he returned, he saw all the cadets in the dining room, sitting at the table, looking very serious. Even Sanyal was looking somewhat nervous.

'Oh shit!' thought Jay, 'the nurse must have complained about the hard-on'

He turned and saw a sight that always made his blood freeze. It was Capt. Jeevan walking along with Sister Rosa Kutty and Matron Sujatha Menon. He was happy and smiling and looked, for all the world, like a happy puppy. Evidently the presence of ladies brought out the inner puppy in Jeevan's case.

He walked up to Jay and remarked heartily to his nursing entourage," So this is the guy..."and beamed at him. Along with blood, other things froze in parts Jay only vaguely registered as part of his body. Things unfroze partially, especially the guilty parts, only when Jeevan continued, still jovially, "The guy who had the six bottles of glucose with the diuretic!" Somehow jovial and Jeevan did not go together.

Evidently the story had been relayed to Jeevan without any mention of the prominent part played by Jay's reproductive parts. Jay was grateful. Punishments were one thing but embarrassment was another thing altogether.

A benevolent gaze such as the one Jeevan was beaming now was even scarier. The benevolent look would have been described by mouse and cadet alike as akin to looking down a python's maw, while being part of a bunch of mice trapped in a corner. The distinction was simple. In both cases you knew something bad was going to happen to you. But in the latter or the 'benevolent gaze' case you just didn't know when...and who'd be first.

Jay joined the rest of the GCs who were, for a change, acting like' gentlemen cadets'. Even Rosa was surprised. Sujatha madam however wished she could get Jeevan to terrorise, maybe even spank some of the other frisky inhabitants of the Officers' ward. Jeevan clapped his hands and announced he was there to conduct the final exams for the service subjects.

Everyone had a look of guilty relief including Sanyal. They had all thought that their conduct had been reported to the IMA adjutant. Conduct unbecoming of future officers and gentlemen though natural in young males, was routinely punishable in the military. This partially explained the presence of Jeevan, the 'GC tamer', in their midst.

They all sat down meekly and wrote, what could at its kindest be considered educated guesswork and as bluntly put by the examiner 'sheer bloody bullshit'. They had no time to study and would not have, even if they had had the time.

Sleep took priority followed closely by food. Still they all wrote as if they were confessing their sins. The exam went off uneventfully except for the time when Jeevan had to swat Sanyal on the head for trying to peek into his neighbour's paper. The neighbour was a I termer and had a different paper. But Sanyal was a creature of habit and copying in exams, cheating etc. had helped him survive thus far. He was lucky that Jeevan was distracted by naughty thoughts of Rosa Kutty and was thus in a benign mood. Otherwise, Jeevan would have been within his rights to report him for cheating, which were grounds for withdrawal from the Academy.

After the exams were over, they all heaved a collective sigh of genuine relief. The relief was undiminished by Rosa Kutty's arrival and announcement of the names of those to get discharged after finishing the paperwork. Jay's name was among those fated for discharge, Sanyal's was not and nobody was surprised.

The spirits were high and the Army tradition of 'dining out' those who were about to leave was maintained. They invited the others from the officer's ward and the duty nurse too, but she wisely declined. She suspected, and rightly so, that she may be on the menu. Jay left the hospital on his own two feet, with all his parts intact and nursing a hangover. This was no small achievement when exiting a military medical establishment.

NOTE

1. Pondy is slang for pornography.

Chapter 21

CABIN CUPBOARDS AND CLEAN SWEEPS

It was that time again. Actually it was referred to as that 'bloody' time again. The time referred to was the 'cabin cupboard' time. Jay had tried to get out of it by taking refuge at the MH but he was back now and right in the thick of it.

Cabin cupboards! The term sounded like the name of some upscale home furnishing company. But the name was as misleading as it was benign-sounding. While it referred to the inspection of the cadet's cabins and cupboards, it was, however, every bit as vicious and demanding as any other parade in the IMA. The use of 'bloody' as a prefix to the term 'cabin cupboard' was truly justified. There were various levels of inspections and punishments handed out that ensured normal life was seriously disrupted.

The punishments were handed out on the spot and noted by the drill JCO or cadet appointment of the company. Apart from that, instant corrections in the form of pushups, rolling and pack parades were enforced by the accompanying pack of cadet appointments.

A cabin cupboard inspection meant the beds were laid out in a particular manner. All clothing and equipment was hung, folded or rolled up and displayed in a prescribed fashion. The floor would be waxed and the furniture oiled. Any deficiency, dysfunctional equipment or item like a cracked mirror or fused bulb would be denoted by a neat placard of specific dimensions, along with the nature of discrepancy and time or date of report. Any lapse in layout or maintenance would invite harsh reprisals.

It all sounded very efficient and as if it had the wellbeing of the cadets at heart. Indeed the ex-NDAs or cadets from military schools found cabin cupboards a minor inconvenience of no consequence. That was probably connected to the fact they were referred to as '*Nangas*'. This referred to their lack of worldly possessions which made tidying up very easy. But the DEs had spent most of their lives with fond parents or devoted servants picking up after them. They were very disturbed by the whole proceedings. Even those who had been staying independently for many years were a bit bewildered. Bachelor pads usually maintain the highest traditions of slovenliness, save for the odd psycho. But when the cabin cupboards were at Academy level, as it was once a year, even the seasoned cadets were rattled.

The inspections were carried out at various levels. The number of levels corresponded to the level of the officer carrying out the final inspection. If it was an Academy level inspection with the IMA commandant coming, then the pain and tension threshold would go up by that much. The inspections would first be carried out by the platoon JUO, then the BUO or CSM, followed by the platoon DSs and then finally the company commander. The battalion commander would also chip in before the commandant came.

These inspecting officials all came on separate days with re-inspections invariably being ordered for no other reason but to keep the cadets on their toes. This meant a better part of the month was spent sleeping on the floor, to avoid disturbing the layout. What made it worse was that many cadets accumulated unsustainable levels of punishments, summary and official.

A few cadets were usually absent during the inspections, being on punishment parades or just avoiding the inspecting authority. This was probably a wise move since, like with most other human interactions, there was the element of personal antipathy involved. Either way, they had to leave their cabins ready and open, with the platoon JUO able to explain their absence. The JUO usually had a lot of explaining to do because he was supposed to have inspected the cabins first.

JUO Akhilesh decided to do just that. Accompanied by two other jobless third term sergeants, he swaggered around the platoon. He did a thorough job and checked under the chests, over the mirrors and other such places normal people don't clean. Many punishments were handed out and supervised by his trusted sergeants. The *lancee* was also one of the victims. The II termers were not exempted and got no leeway from the cabin cupboard. However, they usually got more official than unofficial punishments.

JUO Akhilesh proudly announced, "I give no quarter on line inspection of quarters!" He then laughed heartily at his own weak pun. The others understandably could not share the levity, given their circumstances. Many of the I and II termers were in poses designed to cause major discomfort and pain. The good JUO was actually quite satisfied with the standard of layout and cleanliness but believed a little terror always maintained high standards.

But JUO Akhilesh was not at all happy with the standard of GC Lal Dev's room. To make matters worse Lal Dev was not immediately at hand to feel the full force of the JUO's righteous wrath. That night, Capt Jeevan would be coming to inspect the rooms and he was not a forgiving type. Actually Lal Dev, being a busy cadet with many official punishment parades to complete, had left the tidying up to a I termer. The I termer, not being bound by ties of blood, affection or loyalty, had not felt it imperative to do a sincere job. He had superficially cleaned up and swept things under the rug, literally. In this case it was the dressing table.

That evening Capt. Jeevan came for the inspection. He was coming from a cocktail party and was feeling quite warm and fuzzy inside. He was unusually genial, even cracked a few jokes. Things were going fine till he entered Lal Dev's room and opened his shelf. A dirty sock rolled out. That set the ball rolling. Capt Jeevan politely invited the JUO's attention to the sock and invited him to sniff it. After that was done, much to Akhilesh's distress, Jeevan then asked him to look under the bed. There was some contraband and lots of dirt there. Jeevan was incensed, actually that was an understatement. The next day was the company commander's inspection and this was the sorry state. He awarded the JUO two EDs or extra drill[1] and told him to deal with Lal Dev when he got back from his night report. The good JUO gritted his teeth, snapped a salute and assured him that GC Lal Dev would be shown the error of his ways and life in general.

Thus it was that Lal Dev came back from his night report and was met by a seething JUO Akhilesh Kumar. His first words were, "I've got two bloody EDs because of you and your sloppy bloody cabin, you...you...bloody vagina!" At this stage the good JUO had hesitated because the use of vernacular or obscene abuse was an absolute no no in official communication, no matter how satisfying. It was also because his English was extremely limited. In his native Bihar, English had never come in the way of expressing feelings. After all, there were so many other languages, dialects and forms, from weapons to fists to convey emotion. But he continued gamely, putting his thoughts across in the King or Queen's English, "And now I'm going to make you bleed from every bloody hole you...you...!"

"But sir it's not my fault! I had told that fucking I termer..." started off GC Lal Dev before he realised he had compounded the mistake by admitting he had used a I termer to do the 'dirty work'.

JUO Akhilesh looked at him coldly and said, "Ok, I think we need to have a *puttee* parade[2] just to remind you how to organise your stuff...get into FSMO[3] ... you have two minutes!"

GC Lal Dev sprinted away and was back in a little over two minutes in his FSMO. His tardiness meant pushups for the number of seconds late. He was then given another two minutes and told to change into formal winter mess dress. The process continued for another half hour by which time Lal Dev's room was in shambles. Dresses, boots and other accoutrements were littered everywhere. Lal Dev himself was sweating profusely. But that was not the end of it. He had to fix the room and get it into cabin cupboard state before the company commander's inspection the next day. It was close to dawn by the time he finished. He was not finished however with the I termer who he felt had betrayed him. Lal Dev aka Red God would get his revenge someday.

Jay and Kapoor discreetly watched the fun from their ventilator. Jay remarked to Kapoor "Wonder who's that poor bloody I termer who was supposed to fix Red God's room? Red God's going to screw him thoroughly!"

Kapoor looked sheepishly at Jay and said, "It was me and now I'm feeling guilty that idiot has been screwed so much!"

"Aww screw it, that fucker deserves the *puttee* parade!" said Jay unfeelingly.

While they both knew that Lal Dev would have his revenge, that required time and energy, both of which Green God did not have. So Kapoor was safe for the time being.

The next day Jay and Kapoor decided to go and visit Bhasin and find out what was in the offing. Bhasin being the BCA's understudy usually knew the gossip as also official news. Jay thought it would be a good idea to find out when the cabin cupboard would get over.

They reached his room and found the door latched and the lock hanging on it open. Evidently Bhasin in his hurry had not locked the door properly. Jay and Kapoor decided to go in and wait for him. They entered and found the room spic and span, in perfect cabin cupboard condition.

Kapoor said, "Let's mess up the room a bit!"

"Naww man that would be too cruel a punishment even for an ass like Bhasin... But I would like to find out where he keeps his *pondy* stash...I believe that's how he keeps his dear BCA happy!" said Jay as he rifled through his chest.

"Check under the paper cover at the bottom of his trunk or shelf. That's where he would have hidden it." said Kapoor

"Okaaay! Found it!" said Jay after rummaging around some more. By then all the neatly laid out arrangements had been disrupted. Bhasin had hidden the porn magazine right at the bottom of his trunk. His concern was valid as pornography was prohibited in the Academy. The officers did not take it very seriously and consequently did not hand out serious punishments. But they did confisticate it and it would probably never be seen again till one came back as an instructor. There was, it was rumoured, a fabulous collection of confiscated pornographic material that was circulating between various battalion offices. It was also rumoured that there was an unofficial competition between the various DSs on who would discover or confiscate the maximum *pondies*.

Jay sat down to peruse the girlie magazine, with Kapoor looking over his shoulder. A sigh would escape their lips every now and then. This was the closest they could get to girls. After a few minutes Kapoor said, "Fuck this. I'm going to take a crap!"

Bhasin's cabin was of an older pre-independence vintage and a different design. It had been actually designed for gentlemen and had an attached bathroom. "Don't shit up his cabin cupboard too much!" said Jay and went back to his magazine. After a while Kapoor came out and said, "Dammit! There's no bloody water and the damn flush isn't working!"

"Oh shit!" said Jay

"Oh shit is right, but let Bhasin come and sort it out." said Kapoor

Jay took a quick look inside the bathroom and saw the sign hanging from the flush tank, 'Not working'. "Can't you bloody read dammit. That poor chap is going to be really pissed! As it is, it smells like something died there a week ago."

Kapoor laughed, "*Kya kare yaar* went to shag, *taki dil ka bhoj halka hoga*, but decided to take a crap and *man ka bhoj* also *halka ho gaya*!" [4]

Just as they were wondering what to do they heard the announcement,

'Platoon DS in the line'. "That's bloody torn it! Let's get the hell out of here!" said Jay as both of them got out of there in a hurry. But not before Kapoor added 'brain' to the 'not working' sign.

They later heard that the DS had been outraged by the shambles in Bhasin's room and especially the toxic smell in the bathroom. Bhasin had gotten ten restrictions and the BCA had been hauled up for not keeping his understudy under control. The last Jay and Kapoor saw of Bhasin was as he was straining under the load of two cycles and a double *bajri* pack in front of the BCA's cabin.

Jay met Bhasin later and asked him what happened. All he had to say, with tears of rage, was, "Some bloody bastards fucked up my cabin, man! They crapped in the loo and messed up my trunk and worst of all stole my bloody *pondy*!" Evidently the loss of pornography had been the unkindest cut of all.

Later Jay recovered the *pondy* from Kapoor who was loath to give it up. "But I haven't finished bloody reading it!" he complained.

"A fat lot there is to read in pictorial porn. A *pondy* is meant to be enjoyed not fucking read! It's been specially designed for dumbasses like you who need things in graphic detail so the message gets from your brain to your balls fast! Right now we need to get rid of the evidence!" With that Jay grabbed the magazine and slipped it under Lal Dev's cabin door.

That was not the end of the matter. The magazine had originally belonged to the JUO and Lal Dev had tried to ingratiate himself with the JUO by trying to lend him the same magazine. Lal Dev came to Jay's room later to borrow some packs for *bajri* order, "That JUO fucker has called me...don't know what his bloody problem with me is!" said Green God by way of explanation.

Justice had been served but the cabin cupboard carried on. On the final day, the company commander had for some reason, decided to search Jay's box in detail. He had been slightly amused to find a bottle of brandy and a survival type of knife. "I assume both are for survival!?" he said as he signalled Jay to start rolling and told the CSM to confiscate the contraband.

"You know Red God got screwed again?!" asked Kapoor, "evidently that bloody JUO had owned the *pondy* you slipped under his door, and the BCA recognised it!"

"Oh shit!" said Jay, "that means..."

"Ohhh yeah!" said Kapoor with a satisfied smile, "Now the BCA is convinced it was Red God who screwed up his pet Bhasin's room and is out to screw him!"

"Yep! That's what I call a clean sweep in a cabin cupboard inspection!" agreed a very satisfied Jay.

NOTES

1. EDs or extra drills were the mildest of official punishments and were the only ones generally given to appointments. The term ED was misleading. It was basically lots of strenuous PT in drill dress on the drill square.
2. Puttee parade is one of the dreaded punishments of the academies. It got its name from the puttees that are worn as part of the riding rig/dress and take time being wound. The choice of dresses were wide, PT, riding, drill, weapon training, mess dress summer/winter, service dress, OGs, winter, combat and dungarees. Having to change into different dresses in a short time means the whole room gets messed up.
3. FSMO field service marching order which meant with small pack and ammunition pouches. The dress was simple enough but getting the pack assembled was difficult in two minutes.
4. Literal translation being, 'masturbation unburdens the heart and evacuating the bowels unburdens the mind'.

Chapter 22

THE END OF DARKNESS

The POP (passing out parade) ceremony due to be held in a few days. The whole Academy was decorated, primped and spruced up like some wedding palace. Actually it felt that way with the hordes of visitors and the occasional cute girls, mostly relatives of the GCs passing out. The term 'occasional cute girl' would be misleading; all womankind looked great right then to the GCs. All unofficial punishments were suspended and this edict was strictly enforced. Only the official punishments like restrictions or EDs (extra drills) carried on, but in some discreet corner of the Academy away from the eyes of the visitors. The Academy did not want the parents and guardians to be too shocked or alarmed. Especially by getting a glimpse of what their precious son or brother had been going through the last one and a half years.

Jay was feeling apprehensive to the extreme and was actually lying low. There was a reason for this and we will come back to that later. The only other lot that was lying low were the II termers. They were gleefully biding their time, letting the third termers have their moment in the sun. They could afford to take it easy and wait; after all they would be the kings after the term break.

The I termers had been handed all minor and sundry tasks, from ushering to traffic control assistance. They felt true job satisfaction after all the silly jobs they had been doing for the seniors these past few months. Now they did not have to run around on stupid errands. Every now and then, a III termer would call a I termer and request him (yes really...request, not order) to escort his parents or relatives to some venue.

There were lots of events happening and the I termers did not mind such civil tasks and if there was some young lady in the group, that was a bonus.

Some I termers were in fact roaming around looking for just such assignments.

Jay however, as we know, was lying low or lying doggo, which some say is an even more inert form of trying to be unseen. It had all started 24 hours ago when Jay had gone to the CSD (canteen stores depot)[1] . A III termer had requested them to (there it was again the 'R' word, unheard in the last six months).The weather was good, as was the general morale. The I term was coming to an end and they were going home for some real food and they would be II termers when they came back. They would have more respect and more importantly, time. They had been standing in line at the billing counter and that was when the trouble had started.

There had been a pretty young lady standing in front of them. Jay had been standing quietly when Chib, who was behind him, whispered loudly, "Check out her ass!"

Jay was shocked and terrified the lady might have heard. He hissed at Chib, "Shut the fuck up, you'll get us bloody relegated; she must be some DS's wife!"

Chib was not to be deterred, "Relax man, she looks too young to be an officer's wife, must be some III termer's sister. Let's get an intro and maybe take her out for a coffee."

Before Jay could reply the young woman turned around and smiled at them. Chib pushed Jay aside and introduced himself with a faux gallant attitude and a very genuine leer, "GC Chib at your service ma'am and what's your name?" He also stuck his hand out for further effect and Jay's consternation.

The 'girl' laughed, shook Chib's hand and said, "I'm surprised we haven't met before at any of the company socials."

Chib was very encouraged but Jay had a very nasty sinking feeling. He felt this conversation was taking a turn to someplace full of vengeful senior officers. He was sure this was some senior IMA staff officer's daughter. Chib however did not really care. He was now sure that he was on the inside track to get the girl to have coffee with him and even something off the menu later. He even turned and winked at Jay who was now beginning to perspire hot and cold sweat. Visions of him being tarred, feathered and flogged out of the Academy writhed in front of his hazy vision. It does sound melodramatic, but it was an extremely tricky situation.

The young lady could make out what was happening. She had evidently been down this road before. "You guys must have met my husband; he's an instructor in your company, Capt. Shah."

Jay felt the standard urge all animals have when facing grave danger, 'fight or flight'. He wanted to do both; fight and beat Chib down and then run far away. But the lady must have seen their name tabs and lapel badges, which indicated which term and which company they were from. In fact one could tell a military man's entire life story by looking at his uniform and boots.

Jay was about to fall on his knees and beg forgiveness on behalf of the stupid clot Chib who had tried to hit on the closest thing to a mother there—a DS's wife. Actually there would have been a whole lot less danger trying to hit on Chib's mother than a DS's wife or sister. Chib stood with his hand still outstretched and mouth open...the lips of course were trembling a bit by now.

But Mrs. Shah put them out of their misery and asked them sweetly, "Could you gentlemen please help me carry these bags to the car."

They jumped to it and carried her grocery bags like it was their salvation. They feverently hoped their efficiency would make her revise her opinion. They then mumbled "Good evening" and jogged away as fast as dignity allowed. Mrs. Shah wanted her husband to meet them and thank them for their gallant help. "We'll see you tomorrow at the POP!" She chirped as she waved goodbye.

Jay and Chib looked at each other and wondered what the going rate of retribution was for trying to get over friendly with an instructor's wife. They decided to avoid the POP ceremony by all means possible. Problem was they were detailed as ushers and there was no way they could get away from that. As ushers, they were bound to meet Capt. and Mrs. Shah.

The next day they tried to stay in the background and avoided eye contact with all the guests. Jay felt the emotions of the day as proud parents 'pipped' (pinned on the 2nd Lieutenant's stars) their sons. Everyone was happy, Jay was also almost happy. There was hope on every face, hope of a glorious future. He had very low hopes for his future, especially if Lady Shah had cribbed to her husband.

Jay would be happier when this day came to an end. They would all be off the next day or that very afternoon, and hopefully Capt. Shah would have forgotten the whole incident when they got back after two months. He felt someone tapping his shoulder. He turned around and saw Capt. Shah looking at him in his usual bland way. Jay was initially shocked, then felt defeated and after five seconds, resignation kicked in.

"Where is your buddy, Chib?" asked Capt Shah emotionlessly.

'That's it, the game is up' thought Jay as he stammered a reply, "Yes sir, I think he went to the drill office." which was in the opposite direction. But Shah just said, "I want to thank him and you for helping my wife with her bags at the CSD yesterday".

Jay looked at him feeling relieved. Then Capt. Shah added with a rare smile, "My wife does look kind of young isn't it. Like... some college girl."

Jay also smiled back hesitantly, so she had told him after all. Jay felt something shrivel inside his pants as he forced a smile. He said "Sir, about yesterday..."

But Shah cut him off and said, "It's ok...you know she was Chandigarh's beauty queen, but thanks again, have a good holiday." Clearly he did not want to dwell on the topic. Jay blessed him silently for being a gentleman. Just then his wife came up and said "Hi young man! Where's your friendly friend?"

Jay opened his mouth but before he could say anything she said, "Wish you all the best for your holidays." And they both walked off leaving Jay relieved and feeling strangely emancipated.

Chib came out from behind the tree where he was hiding, and asked him eagerly, "What happened? What did Shah and his wife say?"

Jay looked at him coldly and lied, "They said they will deal with us when we get back from the term break!"

He then walked off with a lighter heart. He also had the mean pleasure of knowing Chib's term break would be somewhat screwed, as he dreaded the thought of getting back and getting punished. Jay decided he could not be so cruel and would tell him the truth before they left but Chib's sins deserved at least one sleepless night. Anyway all that mattered now was the fact that they would be II termers when they got back. Just thinking that gave him the shivers of excitement.

Jay got back to his room and was just about to take a nap before starting to pack when someone knocked on his door. It was the erstwhile JUO Akhilesh Kumar, now 2/Lt (Second Lieutenant) Akhilesh Kumar of the AOC (Army Ordinance Corps). Jay detested the sight of this selfish pig but smiled anyway. There were many seniors who had punished them or made life miserable but this one was a disgusting sadist.

Most of them had done it to teach them valuable lessons on military etiquette or to help pass their tests. Even though the methods had been harsh, it was usually for the greater good. But there were a few who screwed juniors just for the heck of it or for their own selfish needs. This variety was despised by their peers and subordinates. Akhilesh Kumar was one of those, but Jay still smiled and said, "Congratulations, sir!"

After all he was now an officer and what had happened in the past should stay there. But 2/Lt Akhilesh Kumar was not done, "Jay, can you get up that tree and get those bananas down, please". And he turned and looked at his appreciative audience comprising his parents and teen sisters. Obviously they were mighty impressed by the fact that 2/Lt Akhilesh had such command and control over his juniors.

Jay was shocked that this man would want to take bananas growing in the trees belonging to the platoon. Not that anyone ate them, they were mostly taken by the orderlies and sweepers, but it was the thought that counted. Evidently you could make a GC an officer but not necessarily an officer and a gentleman. Akhilesh Kumar had become an officer but definitely not a gentleman.

Jay was about to tell Akhilesh to go to hell, but he saw his family looking on eagerly and didn't have the heart to insult him in front of them. The military code or whatever Jay knew of it dictated respect for your fellow soldier's family. That being so Jay could still not bring himself to climb a tree and get bananas. At least definitely not for a cheap asshole that could well afford to buy them.

"Good afternoon sir and ma'am!" Jay wished Akhilesh's parents cheerfully and then said, "Sir, I'll get the ladder and then get the bananas to your room."

Akhilesh liked the idea and added, "Make it fast. We have to catch a train in an hour!"

Jay gritted his teeth, gave him his most ingratiating smile and said, "Right away sir!"

Jay turned and walked out of the platoon quadrangle and kept walking till he reached the neighbouring platoon, He found Chib's cabin, opened the door, pushed the startled Chib to a side, lay down beside him and went to sleep. He would go back to his cabin after an hour, when Akhilesh and his bloody train had left.

After an hour he woke up and went back to his platoon. The bananas were gone. Evidently Akhilesh had taken them in spite of it all. For a second the mind-boggling image of Akhilesh attacking the tree along with mother and sisters flashed before his eyes. Jay hoped he would never become an officer like that or meet any more such 'gentlemen'. He prayed very hard that he would not have to serve under some mean small-minded bastard like Akhilesh. Thank

God he was going to the ordinance corps which Jay did not intend to join.

"Thank God for officers and gentlemen and ladies like Capt and Mrs Shah." said Jay to Chib who had come to see the 'tree party' for himself. Then he turned to Chib, kicked him in the buttocks and said, "Yes you dumbass, they aren't going to do anything when you get back so go and lech in peace at all the girls you see during the term break!" Jay told Chib who whooped and skipped off.

A few days later Jay was at home catching up on sleep, food and peace. He was at peace as he soaked in the warm glow that comes of basking in your father's pride, mother's love and sister's envy. His father had forced him to wear a blazer and Academy tie to the club. There he had silently sipped beer as his father showed off his son as one would a prize racing horse. He understood how the NDA kids had felt when they had come home on term break. It all seemed so immaterial now.

Jay had been tempted to brag about his time at the IMA, but his sister was not interested and his dad's orderly had pooh-poohed his experiences. I termer training is pretty basic and 'piddly' by any trained soldier's standards. Jay realised there was not much sympathy or awe to be elicited in Army circles so he just concentrated on his mother. She was the only one who was suitably moved by his stories.

Jay thought about all this as he sat in the club. He loosened his tie and let the beer work its magic. His father was there in a trice, "What the hell do you think you are doing?! This is a bloody officer's club; kindly correct your slovenly bloody tie boy!"

But Jay was no longer a I termer, boy or *fauji bachcha*, so he took off the tie and let the beer do the talking, briefly, "I think I'll go home and take a nap...had enough of all this *fauji* bullshit sir!"

Nobody stopped him. After all he had a job and was now his own man.

NOTES

1. CSD is where military personnel get their groceries (minus vegetables and fresh meat) and liquor at a much reduced rate than outside, sometimes even 50 percent less than market rate, especially liquor. But hey, which other organization gives you a tot of rum before ordering you to your likely death.

BOOK III

MAN TO SOLDIER - II TERM

Chapter 23

NEW START WITH OLD FINISH

Jay took a deep breath as he stepped off the train at Dehradun station, even the station smelt better this time around. After two blessed months of term break he was fresh, and ready to start becoming a soldier. The six months of the I term nightmare had made him a man...at least his parents were treating him like one. He was now ready to reap the benefits of all that *ragada* and become something closer to an officer.

The NCO at the reception desk was somewhat polite and actually efficient. Things definitely got better as one became more senior in the army, thought Jay to himself. The other ex I termers gathered around. Hugs were exchanged and sincere questions on lost virginity and fantastical answers on fictional love lives were bantered about. No one talked about mummy's food and family excursions, though that was what most of them had been doing during the term break.

They were brought back to mother earth or specifically the IMA drill square with a thud and a strange cry. The thud was actually a bag hitting the ground next to a III term appointment, the strange cry had emanated from the III termer. This III termer turned out to be an old friend; actually the term 'friend' would be too strong. This was now JUO Iqbal Singh. He gave them a haughty look and growled, "Watch out you fuckers!" before walking off with his cane swinging jauntily.

The new II termers looked at each other and hoped as one that this would not bode evil. Sighting of a demon first off into the Academy, could only mean bad karma for the rest of the term.

They picked up their stuff and went around looking for cabins. They found most of them occupied by the ex-NDAs who had arrived a week earlier. One of them even whistled for Jay thinking he was a I termer. The ex-NDAs having had the experience of three years training were well versed in the fine art of extracting maximum work from juniors. They had thus taken cabins with I termers next to them. That would change of course when the rest of the III termers got back. The I termers were there, wretched dirty-looking creatures, with a familiar beaten look. Jay felt happy looking at them; it reminded him of a faraway past. There would be some suffering; at least it wouldn't be his.

There was a reshuffling of rooms after the rest of the third termers arrived. GC Lal Dev Jay's old nemesis, now a Cpl. (corporal) kicked open Jay's room and succinctly said "Fuck off!"

Jay did some fast thinking and pointed at an ex-NDA II termer's room, "Sir, that room over there is next to a I termer's... also more convenient for snacks and tea since the CQMH's cabin is in that wing..."

Cpl Lal Dev's brain usually ran quite slow, being unable to turn the sharp corners of logical thinking. But in this case, since it was a matter of his comfort and personal welfare, Hari Bhagwan came to a decision fairly quick. "Right!" he said, "pick up my bag and keep it there!"

There was a long silence of the pregnant nature with both sides wondering, 'What the fuck?!'

Finally Hari Bhagwan broke the silence, "Oh sorry...I forgot you're now a II termer...can't ask you for such tasks!"[1]

'This ass will never change' thought Jay as he called an orderly and told him, "*Sahib ka* bag room *mein rakh do*!" and then added for good measure, "*Dus rupiya bhi le lena mat bhulna*!" (Don't forget to take ten rupees from him). Jay's heart was gladdened by the dirty look GC Cpl Hari Bhagwan gave him before walking off to bang on the other II termer's room door.

Jay felt great and perverse happiness at watching the ex-NDA get kicked out of his room. But he also felt sad when he realised he had condemned another I termer to a term of servitude by putting Hari Bhagwan next to him. But then he saw the II termer kick the hapless I termer's door down and ordered him to get out. Jay realised the I termer had been temporarily saved, but the II termer would now be a store that could be readily tapped for anything Red God needed to borrow.

In a spirit of camaraderie Jay volunteered to help the II termer who introduced himself as Ajay Aurora, Hunter Squadron. Jay was a bit taken aback by the added hyphenated title of squadron with the name. He nevertheless shook hands and introduced himself as, "Just Jay of Shankars!" Aurora of Hunter squadron did not get the joke. It was actually quite understandable, if one spends three impressionable years of one's life in a squadron or group, it more or less becomes a part of your identity.

The ex-NDAs were understandably proud of their erstwhile squadrons, having spent three years there with each squadron having its own unique tradition and heritage. They were commonly referred to as '*nangas*'. They did have a marked tendency to bathe in the nude, probably a habit picked up in the NDA where the shower stalls have no partition.

The term *nanga* was also considered by some as an allusion to their propensity to have just the bare minimum of stuff. They had the habit of 'borrowing' any other stuff they required. A direct result of the survival instinct, acquired after three years of training at the NDA. That may not be wholly true, but they genuinely believed that the end justified the means. And of course all means were justified to get the job done and to survive. The DEs learnt these survival lessons but usually too late. For such skills you need to catch them young.

The DEs or direct entry cadets were referred to as the '*dhakkas*', pushed out, or as some would say pushovers. This was because they became officers after eighteen months of training as opposed to the three years in NDA followed by another one year in IMA for the ex-NDAs. Those graduating from the OTA (Officer's Training Academy for short service officers) were referred to as '*jhatkas*' or shake out since they passed out and became officers after just nine months of training.

There was always a latent rivalry between the three types of entries. It tended be a bit intense during cadet days, not so much later. Actually it tended to never entirely fade away especially for the ex-NDAs. The common joke was if any ex-NDA were asked if he knew Shakespeare, he might just ask which squadron he was from.

There was something the other *Dhakka* II termers did not appreciate about their ex-NDA coursemates. The ex-NDAs had spent six terms in NDA and the last term as lords of all they surveyed. The sixth term was very...very fulfilling in terms of moral and social satisfaction. They were the senior-most term in the Academy. They were physically and mentally more robust and attuned to the Military's requirements. There was very little for them to do or achieve in terms of training or testing. And the best part was they had five terms junior to them to boss around, usually unmercifully.

The ex-NDA II termers were now in the IMA. They were junior once more. They had just a handful of I termers subordinate to them. Some, in fact most of the I termers, were older and bigger than the ex-NDAs. That was not all that important, but the problem was that most of the I termers were quite worldly wise and some actually had girlfriends. This made the situation very confusing for the ex-NDAs who had spent most of their teenage hood in a military environment. They were better equipped for survival, but did not have the same moral edge over their juniors as they had had in the NDA.

Unbeknownst to the II termers, ex-NDA or DE the 'moral edge' issue was irking the III termers. They found the reactions of the II termers, specifically the ex-NDA ones 'slow'. They actually had a point, even though it was just a case of history repeating itself.

Many of the ex-NDA II termers had been senior cadet appointments in the NDA and were loath to take orders from their new 'seniors'. Some of the III termers had not been appointments at the NDA but were so now at the IMA. Being a cadet appointment at NDA was good for perks and the ego, not much else. But being an appointment at IMA meant you went up in merit and could get the regiment or arm of your choice. Thus the competition was that much more and in earnest. Cadets who lay low through three years of the NDA now woke up and started pushing their potential.

The fact that they were now gentlemen cadets and not just plain cadets did not change the dynamics, or so they thought. Matters came to a head sooner than expected. One fine rainy evening, GC Sgt. Rajesh Pandey called a II termer, actually he just whistled for him. Pandey had been a lancee before, and was generally well liked by the DE II termers; they had seen his decent side as a II termer lancee. The II termer whistled for, GC Aurora, however was an ex-NDA and did not know or care about Sgt. Pandey's past achievements. He had been a CSM at the NDA and thus considered himself above lower mortals like DE II or even III termers...especially the likes of Sgt. Pandey. He thus ignored Pandey.

Pandey was stung by the blatant disrespect shown by the II termer. He too did not care about GC Aurora's past achievements. What he did care about was losing face. This was the relatively calm period after games. Dusk is when normal humans took time to take a deep breath and reflect on the day. In the case of the IMA it was just a brief lull before getting ready for the evening dinner or roll call and other such fall-in's. This was also the period when many juniors aka I and II termers were roaming around. The ignoring of Pandey's inviting whistle would soon become the joke of the day. Pandey realised this and was doubly infuriated.

Pandey decided to call the II termer whose name he did not know directly, "Hey you! You...that fucker with the NDA blazer, come here!"

The man in the NDA blazer, GC Aurora, calmly kept walking. He knew this could get ugly. Aurora also knew from experience that he had deniability, since he had not been called by name and had assiduously avoided eye contact. Thus GC Aurora, ex-NDA of Hunter squadron strode tall into the bathroom area and vanished into the gloom of the unlit corridor thereof. Increasingly strident screams from Pandey followed him. Aurora paused in the gloom to see if there was any pursuit.

GC Jay happened to walk by at just that very moment when Aurora was disappearing beyond the bathroom corridor bend. Jay was at peace with the Academy, the first week of II term life and the world in general. He no longer had to look over his shoulder and worry about being punished, tasked or generally harassed. Just then Pandey yelled out to him, "Jay get your ass here!"

For a second Jay was transported back to the I term. It's an old Army saying that 'a senior may forget he's senior but a junior must never forget he is junior'. Revelling in his status as a II termer and relatively senior in the Academy, Jay had for a minute lost sight of the aforementioned axiom. Pandey's voice reminded him that he was still a junior, and he still had people senior to him in the Academy and said people were very close right then.

Pandey was generally considered a decent chap so Jay jogged up to him as a courtesy and asked, "What's up sir, someone flicked your boots?" This was a common occurrence when boots were left outside for the orderlies to polish, usually sparking a chain reaction of boot lifting.

"Run and catch that bloody II termer! There, there, there!" pointed Pandey in his agitation.

Jay liked Pandey, but the unnamed II termer was a coursemate plus it was bad form being asked to perform such tasks. He had a bad feeling about this and tried to stall.

"Where sir? Which way?" He asked with a faux innocent and confused look.

Pandey wasn't buying it and said, "Stop giving me those fucking lost I termer type looks! That bugger's gone towards the bathroom. Go and bloody well find him fast!"

Jay tried one last time, "Sir any idea who...?"

"Will you stop fucking dilly dallying and get a bloody move on! It's a fucking ex-NDA in their *chutiya* blazer!" said Pandey.

Jay shuffled off unwillingly and turned the corner into the bathroom corridor. It was so called because there were only bathrooms in that dark and smelly corridor. Generations of juniors had hidden out there till the coast was clear. Jay saw Aurora the moment he stepped into the corridor. It was an awkward moment.

Jay cleared his throat and said, "Uhh... I think that III termer is looking for you..."

There was another awkward silence before Aurora deigned to reply, "I know."

Jay could not appreciate the brevity and clarity of the reply since he was in a moral dilemma. "But he's calling you..." he tried again.

"I know!" came the terse reply again.

Jay realised this horse was not going to the water or forced to drink anytime soon, so he tried one last time. "What do you wanna do?"

"I'm going to my cabin. That bloody III termer can go fuck himself!" came the informative answer.

The dilemma was akin to an age old one, the choice between wife and mother. Jay liked Pandey, and Aurora he did not know well. But betraying a coursemate was considered second only to incest in the Academy's order of precedence of unpardonable sins.

Jay took a longish walk around the platoon and came back with a suitably worried look. "Sir I could not find him...was it something urgent?"

"Fucker...don't give me that bullshit! You get on your hands down!" said Pandey none too kindly.

This type of reaction was uncharacteristic of Pandey and showed how pissed off he was. Just then Pandey saw the CSM and went off to crib to him. Jay overheard a part of the conversation before he, feeling blessedly neglected, slunk off. The gist of the exchange was simple. Pandey went and cried on the CSM's shoulder who was not very sympathetic. Pandey then put forth his case more forcefully and put the CSM also in the 'choose between mother and wife' quandary. But it was obvious something serious was going to happen and happen very soon.

Matters came to a head when the BCA called a II termer who by unhappy coincidence happened to be Aurora. Aurora having been emboldened by his encounter with a fairly lowly Sergeant decided to up the stakes and push his luck with a battalion cadet appointment, the BCA.

The BCA was fairly high up the totem pole and one of the key cadets the Academy adjutant and battalion commanders relied on. Aurora had pushed the wrong button and not only upped the ante but ensured all the II termers of the battalion were going to get 'a bamboo up theirs...'

In fact those were the very words used by the BCA while discussing the attitude problem of the II termers. The actual issue was based on old military wisdom which insisted that defiance of authority is infectious. The III termers's authority was very tenuous, based mostly on tradition. The last thing they needed was the I termers getting ideas. A large scale revolt was every military man's nightmare.

Aurora, of course, was congratulated by the rest of his NDA compatriots on his courage of convictions. The rest of the DE II termers could not fully appreciate the attitude. It had been not-too-subtly conveyed that the ex-NDAs had finished their quota of '*ragda*'...unlike the DEs. Thus implying the DEs had some more toughening up due. This sentiment, understandably, did not go down too well with the DEs. But like other immutable laws of nature one could choose one's friends, maybe even wife, but not coursemates.

The monsoons had begun and the heat had abated. The atmosphere at the Academy had however reached fever pitch. The ex-NDA II termers would have staged a coup, if they could have. The III termers knew no such thing could happen and that the II termers wanted to put themselves on par with their seniors was unthinkable. Such impertinence could not go unpunished. All the third termers were unanimous in their opinion-the II termers would have to be 'sorted out' and their doubts removed, at the earliest. The DE II and I termers would be collateral damage, but that was just too bad. Course spirit had to be maintained and the protocol of juniors feeling the pain of their seniors's suffering had to be maintained.

The III termers decided against having a central 'doubt-removing' fuck fall-in. It would be company-wise. They also decided against enlisting the aid of the officers. After all if you want to show who's the boss of the schoolyard you don't call your dad to back you up. Then dad would become the boss...if you get the reasoning. The most irksome of the doubts were the sullen looks and dumb insolence. These were usually precursors of larger acts of defiance.

Most despotic regimes around the world carry out routine 'sorting out' exercises just to weed out precisely these doubts. And so it came to pass one pleasant Saturday evening, the order went out for all II termers to report in *bajri* order to their respective platoon JUOs after dinner. There was hectic manoeuvring and lobbying. Favours of all sorts were called in, old relationships reviewed. But the third termers were unmoved. No II termer was excused.

Jay and the other II termers of the platoon fell in in front of the JUO's room. GC Sgt. Pandey was there with a very self-righteous smirk that approached the obscene. The JUO came out after about half an hour. The delay was calculated. The II termers had in the meantime, grown restive and were chafing at this time-wasting. They were all toughened up to quite an extent. While no one liked a fuck fall in, it no longer held the same terror as it did in I term. The III termers knew that and focussed on the time wasting part.

The plan was simple. First let the II termers exhibit some more of attitude, so they would know the exact reason they were being screwed, thus removing any doubt of the 'doubt' they were being taught a lesson for. Then they would be screwed till they cursed the origin of the doubts. Then finally the II termers would realise that though the fuck fall in would not be really painful enough per se to show them the error of their ways, they would waste a lot of time in acute discomfort. The wastage of precious time was what would pinch them the most.

Accordingly the JUO kicked off the proceedings with a long winded ML. The ML is likened to a power point presentation. The official definition of this was something on the lines of—'the wisdom that comes out of the presenter's mouth and goes through the ears of the audience without touching the minds of either'. This was an apt description of the situation during an ML but the key again was wastage of time. Then of course, there was the physical part.

The II termers were doing pushups, actually an advanced version of pushups with *bajri* packs. This, in all fairness, was actually the fault of the II termers. When the order was given to keep the packs to one side and get on their hands down, Aurora went into a 'Rambo' type mood and with a snigger decided to get on his hands down with the pack. The message Aurora wanted to convey, 'I'm a tough soldier and I can take anything you pussies can give me!' was understood by all present.

One section of the audience, notably the III termer appointments got the message and did not take it well. This was again a challenge to their creativity in organising memorable 'fuck' fall in's. The rest of the audience, comprising II termers and the few hidden I termers were horrified. Jay heard someone next to him whisper, "Guess now we know Aurora prefers anal gratification...without lubrication."

The III termers picked up the gauntlet. Those just watching on the sidelines moved up to assist and supervise, thus making the whole event personal. A personalised fuck fall in was always to be avoided since it involved intimate supervision and there was very little chance of 'shamming it out'[2]. There were amazing theoretical similarities with the actual carnal act, but no one had time for the deeper philosophical implications.

Sgt. Pandey had promptly stepped into the breach and smugly suggested that the II termers not bother to take off their packs since some of them seemed to be so fond of it. The II termers smiled bravely and some like Jay, just grimaced and attempted to pass it off as a smile. Course spirit aside, the DEs did not want to be seen lagging in bravado. And down they went on their hands, with the twenty kilo gravel-laden packs.

It was finally around 2 A.M. when the II and third termers decided to call a truce. The third termers were feeling sleepy. They were running out of ideas. The II termers were about to enter the state of limbo where they would cease to feel pain and just do things automatically. Thus far no one had collapsed except Sandy Ahlawat, but the very fact that he had lasted three hours was an indication of the II term course resolve. The II termers had got the message. At least all participants in the fuck fall in so hoped, both the giving and the receiving end.

For the sake of appearances Sgt Pandey gave one last ML which lasted about five minutes before one of his coursemates told him irritably to wrap it up. The II termers were asked to break off. Some of them shouted hip hip hurray and a few even whistled. The BCA who was passing by invited the II termers back for an additional round of pushups and they acquiesced.

Bravado was the mood of the night as far as the II termers were concerned. But even the BCA lost interest after thirty pushups. In fact he also belted out fifty with them, just to show them he was as tough. The BCA's efforts were loudly applauded and cheered by the watching I and II termers. After all, high spirits after a solid five-hour *ragda* indicated good soldierly qualities and high adrenalin levels. The BCA just smiled said, "Get lost you silly buggers!"

The II termers felt exhilarated as if they had passed a test for manhood. The I termers didn't know what the hell was happening and why but they all yelled, "Good night sir!"

The III termers also smiled as they went back to their beds. The ice had been broken, albeit violently. There was now mutual self-respect between the II and third termers. They had as most animals do, marked out their territory, their lines of respect as it was in this case. Much as Jay and the other juniors wanted, this story could not end with 'they lived happily ever after' or even 'slept peacefully ever after'.

The II termers were filthy, covered in mud and soaking wet. At some stage they had been told to take off their uniforms and now their undies were indistinguishable from the rest of them, owing to the quantity of mud inside and out. The first order of business was to have a bath. The CSM promptly switched off the mains from the fuse box.

The II termers cleaned themselves up in the dark, quietly washing what they could reach or feel. It was the closest humans could come to actually licking their wounds.

Kapoor remarked, "Tomorrow I may have to use sand paper instead of toilet paper...bloody mud everywhere..."

And then it was desperately quiet, as the II termers attempted to catch at least three hours of sleep before the next day dawned, most probably a hostile one at that. The II termers's civil disobedience movement had officially been suspended, indefinitely. The aims had been achieved, though Jay was still not sure what the bloody aim was...maybe someday when he became a III termer he would understand.

NOTES

1. II termers were not given minor duties or chores, unlike I termers who were. They also could not be punished by ordinary III termers, only the appointments. Though technically, any third termer could tell his course mate, an appointment, to do the needful.
2. Shamming it out referred to the practice of looking like one is working very hard or sincerely while doing the opposite.

Chapter 24

Do Unto Others

The importance of having juniors has been stressed in this tale quite often. It had been stressed on Jay too and in all manners of stress at that. As Jay understood it, there was no point being senior if there was nobody junior to you. So in essence, the key feature of feeling senior was having someone junior to you.

Thankfully, God and country had placed I termers there in the Academy. They played a vital role in providing III termers bed tea. They were a ready source of essential items that could be borrowed in times of emergency or just for the heck of it. Most importantly, they provided the II termers with a sense of self-worth and the all-important ego boost. The II termers who looked for the last two returns from the I termers were usually considered the dregs of the Academy. But it takes all sorts and nobody said a military community was utopian.

A military academy can at best be described as an organised version of hell with discipline and lots of PT. Even Satan would have quit from such a hell. Such environments and times of stress brought out the best and the worst in humanity. They say if you want to find out the character of a man, send him to the Army. Some characters should never be found in the Army. Sandy was one of them.

GC Sandy Ahlawat had barely made it to II term. He had been held back in the term break and had passed out only after the PT SM had taken a personal interest. Popular rumour had it that Sandy had improved his 'chin up' by keeping his chin up perpetually in the 'arse kissing' pose.

Yes, the GC Sandy Ahlawat of II term was a new man and a bigger ass and if his coursemates were to be believed, literally. He had also turned out to be a bully. He had never had the chance to boss over anyone at home, thanks to the fact he had two older bossy sisters and an equally bossy mother and mild henpecked dad. He had naturally assumed that men were supposed to be ordered around, till he came to II term. 'Power corrupts, absolute power corrupts absolutely' or so goes the old saw. But it is also true that a little bit of sudden power can also corrupt thoroughly. So it was with Sandy.

The II termers generally ignored the I termers as they were too busy with their own routines. In any case the poor I termers were usually too busy running errands for the III termers and had a pathetic air about them. Their discipline and welfare such as it was, was looked after by the II termer lance corporal or lancee as he was popularly referred to. The discipline of the I termers was the responsibility of the JUO of the respective platoon and the CSM. In theory, any senior could discipline or punish any junior. But in principle, nobody, apart from the cadet appointments, had the time to bother about disciplining juniors in the Academy. The curriculum was packed with exhausting training events as well as exams and practical tests. The only occasion any senior other than the appointments bothered to discipline a junior was when it became personal or if that particular senior was mentally and physically disturbed. There were a few of those cases.

Sandy was not really a psycho but for the first time in his life he had someone he could actually order around. He could strut and preen and talk without being laughed at. At home he had been mocked by his sisters and then by his coursemates. The only person who had ever taken him seriously was his mother, but now he had juniors and he would make sure they feared and respected him. He tried on several occasions to get them to assemble in front of his cabin but was always thwarted. The moment they fell-in, some III termer would call them or the lancee would appear on the scene and whisk them away.

But Sandy got his chance soon enough. He overheard the JUO telling the CSM about the necessity of resuming the night PT to improve the physical fitness standard of the I termers. Their performance in the PT tests would go towards the inter company championship points. Sandy jumped at the opportunity and after dinner, just before lights out slithered up to the JUO's cabin and said, "Sir I would like to volunteer to train the I termers!"

The CSM was also present in the JUO's room and their first reaction was to burst out laughing.

"What the fucks are you going to train them in, you silly ass?" asked the CSM succinctly.

Sandy's performance on PT tests was common knowledge. In fact his name was synonymous with being a weakling. Sandy was aware of this and also of the fact that any minute now these two III termers would humiliate him and 'have' him. His brain worked overtime and he said with a serious expression, "Sir, I now know how to improve my weaknesses and it would also help me improve with them..."

The JUO and CSM looked at each other. The CSM, who was a fair man, decided to give Sandy a chance. The fact that the CSM had seen Sandy's sister may have tilted the scales in his favour. "Ok, but just the I termers in your platoon-and no fingering them unnecessarily, just bloody PT! Got it?" asked the CSM.

"Yessir, yessir! I will make sure they all pass their tests...sir" panted the excited Sandy looking like an excited puppy.

"Alright, but the JUO will keep an eye on you, give him the report!" said the CSM, "Now fuck off!"

Sandy turned to the JUO with a grin that would have been considered lascivious on the face of a female of any species. But on Sandy's sappy mug it just looked smarmy. The JUO looked at him stone-faced. He had not seen Sandy's sister and was thus not at all impressed. He considered Sandy a leech unfit to wear a uniform, leave alone become an officer.

The JUO said curtly, "Make sure you give me a report before and after the night PT fall-in." The JUO's tone indicated dismissal, but after a pause when the JUO turned around he saw Sandy still standing at the doorway with a goofy grin. He smiled and said, "Kindly get the fuck out of my cabin! Are you expecting a bloody goodnight kiss?"

Sandy, confused by the smile, made the mistake of smiling back. Fifty pushups later, Sandy backed out hurriedly and rushed to his cabin. He slept with a smile even though his arms and shoulders were aching. He, after all, finally had something to gloat about.

That night he dreamt he was on a throne, his mom next to him while unending lines of I termers bowed down before him. The JUO and the CSM were slaves in chains of course, in this dream.

The very next night itself Sandy effected his plans without telling his coursemates. The other II termers wondered why Sandy was swaggering. They could not come up with any clues why a 'Sad Sack' like Sandy would have any reason to stand tall and proud. Only Aurora guessed and told Jay who was sitting next to him at the dining table.

"That silly bugger is going to finger the poor I termers today." Aurora remarked, matter-of-factly.

"What?" said Jay "Sandy?" The thought of Sandy exerting himself was unbelievable.

"You'd be surprised how even the biggest loser can get an ego boost from suppressing his juniors," Aurora said and continued, "And shago[1] DEs are usually the ones who like screwing their juniors."

Jay supposed Aurora knew what he was talking about. He had only spent six months suffering under such bloody-minded bastards, while Aurora had probably seen many more such types. After three years at the NDA he could spot or maybe even smell sadists from a distance. Jay informed Kapoor who was sitting next to him of Sandy's likely plans. Kapoor duly passed on the information to the II termer sitting next to him. Within a few minutes the entire Bhagat battalion mess dining hall knew about Sandy's plans. For his part, Sandy could not understand why some people were looking at him and laughing while most of the other II termers were giving him looks of undisguised contempt.

A I termer went up to the head of the central dining table and announced that all I termers of Singarh company 3rd platoon would fall-in, in front of the JUO's cabin at 2030 hours. The JUO had the I termers fall-in in front of his room after dinner and then told them to report to GC Sandeep Ahlawat. It had been Sandy's request that they come to the JUO first. He had not been sure if the I termers would come directly to him.

Sandy was ready and waiting in his dressing gown, looking for all the world like Hugh Hefner about to have a party with his playmates. Only this was not the Playboy mansion but the dimly-lit, dew-soaked lawn of the platoon quadrangle. It had taken Sandy some time to decide what dress to wear. Uniform would have been too formal; he wanted to show the I termers that he was cool. PT dress would indicate that he was going to work as hard as them and drag him down to their level. Other dresses like riding rig or IMA blazer would have been out of place and in any case he did not want the JUO to get suspicious. Sandy resolved this quandary by wearing the prescribed night dress, pyjamas with the mandatory dressing gown.

The I termers assembled in front of Sandy's room. One of them knocked on the door and nervously said, "All in, sir..."

Sandy, who had been waiting eagerly by the door, waited for a minute before answering, "Wait you bloody buggers!" After another pause he then opened the door swaggered out and gave a theatrical yawn.

"Ok so you're the buggers..." said Sandy.

There was an uncomfortable silence as Sandy was at a loss for words while the I termers looked confused. He then remembered and said, "Ok, give me the report!"

There was some more uncomfortable silence as the I termers looked at each other. There were only five of them so giving a report seemed kind of redundant. But Sandy had to give a report to the JUO and his vague military knowledge told him he could only give a report if *he* had been given one. Finally one of the smarter I termers realised the only way to get this farce over with soon was to do whatever this clot wanted.

The sight of the pathetic looking I termers in their dirty PT whites would have moved the hard-hearted to laughter and maybe mothers to tears. Most II termers would have told them to get lost and go to sleep. Sandy was also somewhat inclined to let them go. But this was one of the finest moments in life. He wished his mother had been here to see it. He also wished his bossy sisters were in the fall-in along with the I termers.

He then snapped out of his reverie and remembered that he had to report to the JUO.

"Ok *savdhan*!(Attention) I'm going to give the report to the JUO, " said Sandy and swished off in his flowery silk dressing gown (his sisters' wildly inappropriate parting gift).

He knocked on the JUO's door and said in the same tone as the I termer, "Sir, five I termers here for night PT!"

The JUO opened the door, looked at Sandy and asked, "What the hell do you want?!"

"Uhh sir, the I termers are here sir... for the night PT sir..." stammered Sandy.

The JUO gave him a look usually reserved for overly-insistent beggars, "Ok just half an hour and just PT got it? Now fuck off!" With that the JUO slammed the door and went back to swotting from his tactics precis.

Sandy went back to the I termers who were trying hard to suppress their grins. They had heard it all and realised Sandy had a very limited mandate. Sandy saw the look on their faces and was outraged. Being dismissed summarily by the JUO in front of the I termers was bad enough but being openly disrespected by them was just too much. He did not notice Kapoor, Aurora and Jay laughing in the shadows.

"Ok hands down you buggers!" said Sandy in his gruffest voice.

The I termers knew what Sandy wanted them to do but instead of getting into pushup position simply bent down and put their hands on the ground.

"No, no you silly buggers get into pushup position...like this!" said Sandy with a tinge of exasperation as he got into pushup position himself.

The I termers realised they had struck gold and 'doped' it out with a vengeance. The next ten minutes Sandy ran from man to man as he attempted to genuinely teach them how to do pushups. Every time he corrected one I termer, another would do something silly on purpose and Sandy would demonstrate the correct technique. After about 15 minutes of this Sandy was sweating more than the I termers.

Jay and Kapoor watched the fun from the shadows. They were mildly concerned that Sandy would make an ass of himself and the DE II termers a laughing stock. Aurora was waiting for just such a thing to happen.

Then things turned nasty. Sandy noticed one of the I termers smiling. This was GC Sidhu. It was the same chap who had displayed initiative by knocking on his door and then giving him a report. He was smart kid from a famous military school called the 'Georgians'. He thus had the advantage of having served in a military environment through his formative years and could thus smell out weakness and ineptitude. Both qualities GC Sandy had a surfeit of. He turned red, puffed up and said to Sidhu, "Get up fucker"

GC Sidhu got up slowly. Just slow enough to irritate but not slow enough to make it look deliberate. Sandy started a tirade which did not make much sense. The I termers started smiling openly. This infuriated Sandy even more and he snapped, "Get on your hands down *bahinchod*!"

This was exactly the opening the I termers had been waiting for. Abuse in vernacular with other cadets especially juniors was considered unofficerlike. Terms of abuse implying carnal knowledge with mother or sister were to be avoided unless there were cordial relations between abuser and abused, so to speak. This was not the case between Sandy and the I termers. The I termers got up and looked at Sidhu who looked relieved. He had not expected this stage to come about so soon. He then said, "Sir please don't call me *bahinchod*!"

This had the desired effect and Sandy totally lost it. He grabbed Sidhu by his collar and said, "Fuck you!" That was a mistake. Now Sandy had lost all moral ground. He had used a cuss word that was so not recommended. Then he had compounded the mistake by grabbing Sidhu's collar which officially came under the offence of manhandling.

Sidhu knew he was at an advantage. He saw an opportunity and went for it. He immediately grabbed Sandy's hand and forced it off his collar. Sidhu had been a boxer and an athlete in school and college. He was stronger and fitter than Sandy would ever be. There was no way he was going to let a weak excuse of a senior like Sandy push him around. Especially now that it was established in front of witnesses that he had been manhandled and abused.

Jay and Kapoor saw their worst expectations come to life in the form of a sordid tableau in the platoon quadrangle. The ramifications were too terrible to contemplate. The DE II termers of Singarh company would become the laughing stock of the Academy. The I termers would start 'showing doubts' aka questioning orders. Their defiance would have to be crushed.

Sorting out the I termers would be mandatory and time-consuming. It would also mean some of the third termers would be deprived of the I termers' services. That would piss them off and they would take it out on the II termers. At some stage the officers would also get involved which would be a disaster of very painful proportions. And all this because the silly ass Sandy wanted to prop up his ego and feel like the boss he would never be.

Aurora reacted faster than the others. His intentions were to establish ascendancy over the I termers, who would then fear him even when he became a III termer. It had begun to rain as if even God was weeping over Sandy's incompetence. He walked briskly over to the scene which had frozen with Sidhu holding Sandy's hand while Sandy stood there with his mouth open. Sandy's brain had temporarily disowned his tongue and he stood there, speechless.

Aurora quickly took charge of the situation. He had handled juniors before. He was proud of the fact that they had considered him a bastard in NDA. He told the I termers to get into *savdhan*. That was a legal order that no junior could refuse. He then told Sandy to stand back and ordered Sidhu on his hands down. Sidhu realised this was different animal but tried hard one last time for all I 'termerkind' by acting dumb. "What sir...?" he asked.

But Aurora was having none of it. 'Always target the leader first' was one of his dictums when handling restive subordinates. After all he had had the benefit of three years at NDA with scores of unfortunate juniors.

Aurora just snarled, "Shut the fuck up and get on your bloody hands down!" Sidhu appreciated clear and concise orders especially ones he could find no legal fault with. Sandy also shut his mouth since he correctly assumed the first part of the order was meant for him too. Within a few minutes the I termers were sweating and doing the PT routine without too much guidance except the occasional threat. Sandy stood there, looking and feeling very foolish. He knew he had lost moral ascendancy over the I termers forever. All his plans of an ego boost from the I termers were dissolving along with their sweat in the grass. He just hoped his other coursemates did not get to hear of this.

Jay and Kapoor looked at each other and could recognise, even in the shadows, their identical expressions of horror. They were more demoralised by the sight of Aurora than had been by Sandy's tragi-comic drama. Sandy's ineptness would now be retold throughout the Bhagat battalion as the DE GC's incapability. They racked their brains and then the solution came to Jay.

He went up to the nearest III termer's cabin and hammered on the door before scampering off. The door opened within seconds. It was GC Sgt. Lal Dev who had thought it was the duty officer and had thus rushed to open the door. The anxiety and apprehension soon changed to anger and extreme irritation.

Lal Dev saw in front of him I termers of the platoon and the two II termers, Sandy and Aurora. Both were opposites and their collaboration in a I termer fuck fall-in smelt suspicious. He had been specifically banned from using I termers to run his errands and here were two II termers misusing them! He decided to join the party.

The I termers were soon despatched on chores and errands. Mostly things like fixing Hari Bhagwan's packs, borrowing stuff for his assignments or carrying messages. Once Lal Dev was satisfied that all the I termers were gainfully occupied, he turned his attention to the II termers. They were in different attitudes.

Sandy was preoccupied trying to think up a story he could tell the JUO for having failed to administer night PT. Aurora on the other hand, was so cocky that even Lal Dev who was dense enough to miss ordinary human emotions, noticed it. That was a big mistake and Lal Dev settled down to give them an ML. Fortunately for them the JUO poked his head out after a few minutes and ended the ML with a penetrating question about the I termers's welfare. The JUO's question was answered by a long-winded monologue by Lal Dev about how he had rescued them etc etc.

The JUO got the drift of Lal Dev's rambling in 'Hinglish'[2]. He turned to Aurora whom he mistakenly identified as the offender and asked coldly, "I thought I told you fuckers not to screw the I termers...just make them do night PT?"

Aurora confident that he had been doing public service said, "I was just helping out Sandy... who was not able to control the I termers...Sir!" He added humbly when he realised the JUO was not feeling charitable. He was right.

"Tomorrow night both of you bloody well come with the I termers..." said the JUO.

Sandy added fuel to the fire by asking, "Sir what dress should we come in?"It was a genuine but wholly unnecessary question, but Sandy's luck was ill-starred that night and the question ill-timed. The JUO, who had almost shut his cabin door, turned around and came out. There was a pregnant pause while the JUO was lost in contemplation. Aurora wanted to pull Sandy's and his own hair out. He rightly feared the worst, again from long experience. Then he said, "Come in *bajri* order gentlemen!" and with that he slammed the door, decisively.

The JUO had shut the door fast because he wanted to make sure Sandy did not ask any other questions that might lead to his death. He had noticed the murderous look Aurora had given Sandy. In fact Sandy had been about to ask if they should come in triple or double pack. That would have surely pushed Aurora over the edge.

The murderous look was repeated when Aurora bumped into Jay and Kapoor in the shadows. The duo had been loath to miss any of the proceedings and stayed inspite of the danger. The three looked at each other and then they all went to their respective cabins and had classically Freudian dreams.

Jay dreamt of saving I termers who then turned into sexy mermaids who were eager to show their gratitude at having been saved from the evil witch and goblin who looked very much like Aurora and Sandy. Kapoor dreamt he was riding a horse that looked like Lal Dev while he rescued small kids from the evil ogre Aurora.

Aurora dreamt he was emperor, with the JUO as one of his slaves deputed to sodomise Sandy to death. He smiled as he slept. He often had homoerotic and sado-machoistic dreams like that. Sandy dreamt he was wearing a skirt and playing with his sisters. They had dolls that looked like the I termers and his sisters wouldn't let him play with them. He sniffled in his sleep.

NOTES

1. Shago is short for 'shagged out', a euphemism for being tired or weak. It is considered a derogatory term.
2. Hinglish is a common reference to English with Hindi or Hindustani terms. It was also used to refer to the kind of English spoken by people like Lal Dev, who used more Hindustani words than English in their sentences while trying to speak English.

Chapter 25

SWIM SINK OR JUMP

Every officer is expected to know swimming. Swimming is thus a skill strongly recommended to anyone who wishes to join the Military and definitely for a leader of men. The IMA also vigorously agrees with that line of thinking and made swimming fifty meters, unaided if you please, a mandatory test. It came under the 'must know' category of skills to be learnt before becoming an officer.

Swimming is considered fun in most circles, at least where it is voluntary. But in the Academy it was considered a misery and or at the very least a buggery. Buggery is when someone has fun and misery is when nobody has fun. There were many who knew how to swim and looked forward to these classes—at the beginning. Everyone hated these classes later.

The primary approach of any military training institution was to keep everyone gainfully engaged. Imparting military skills came later and that too under duress. There was probably a grand design and a method behind all the madness. But it took, on an average about twenty years in the Army for a person to fully appreciate it.

The swimming class was on a Tuesday, in the two periods before lunch. It was just the right time to have a soothing turn around the pool and work up an appetite. But there were some who had lost their appetite on Saturday itself, when they saw swimming on the weekly timetable.

Swimming classes had been part of the curriculum in the first term, but not taken too seriously. The emphasis was on separating the dead sinkers from the floaters and giving basic swimming lessons to those who needed it. Sometimes, it was the only time when water touched many a I termer's skin so thoroughly. Personal hygiene was very low on the list of priorities. Being on time, keeping out of seniors' way and sleeping was the main focus.

The changing room was actually a kind of refuge for some I termers. Even the toughest PT instructors did not go into the swimming pool changing room. The stench of unwashed humanity, sweaty clothes and not to forget, socks, made the combination quite powerful and repulsive. It would have killed anyone without experience with sons or brothers. But even Sandy, who had lived in a frilly, sweet-smelling feminine environment all his life had actually hidden there and even slept. That exemplified the depths humanity could go to in the quest for survival.

Those who knew swimming were given other things to keep them busy like push-ups and rolling. Seeing this, Jay had joined the beginners and pretended to be scared of getting into the pool. Nothing amuses a swimmer more than watching a non-swimmer in distress. The instructor's baser instincts were appeased and they generally left the hydrophobic non-swimmers alone and concentrated on the others. This meant that they could stay in the pool splashing around dramatically, pretending to learn how to swim. Swimming in the first term was just a minor buggery. Some still managed to have some fun.

In the II term, the dynamics totally changed. It was a mandatory test and that meant one could not hope to become a III termer or even an officer, if one did not pass the 50 meter swimming test. Poor results meant problems for everyone, including the instructors. They took such things very seriously. Present today was an instructor who was a paratrooper. He cut a very dashing and chilling sight. This sort of effect one usually experienced while observing some feline predator in the wild on TV. But this animal was here...amongst the prey.

The officer was here for the same reason predators usually make an appearance; to cull the weaklings. His potential prey knew it too and tried to hide in the changing room.

The PT instructors came into the changing room and herded the unwilling II termers out to the swimming pool. They used their whistles and finally even their whistle straps, till all the II termers had fallen in. They, in skimpy maroon IMA swimming trunks, stood to attention while Para commando Capt. Balhara looked them over.

He said softly, in fact very softly, "Swimmers and non-swimmers..." It was menacing enough to get them all activated.

The PT instructors sprang into action, the II termers jumped and scrambled, a clipboard appeared, and soon the swimmers and non-swimmers were segregated. Jay found himself standing with the non-swimmers, and had a bad feeling about this. When he saw Capt. Balhara look thoughtfully at the deep side of the pool and having a brief aside with the senior PT instructor, Jay decided to turn honest. He tried sneaking up and joining the swimmers but was halted in his tracks by Balhara's voice.

"Where the fucks do you think you're going, son?" he asked.

Jay stammered, "Sir, I am a swimmer..."

There was a pause while Balhara assessed the situation and came to the right conclusion. "Start rolling around the pool. Maybe that will help you make up your bloody mind faster in the future." He then ignored Jay and went back to his discussions with the PT instructors.

By the time Jay had completed rolling around the swimming pool, his back looked like an alligator's. It was deeply ridged because of the chequered anti-slip pattern of the flooring around the pool. In the meantime the other II termer swimmers had been lined up behind the starting boards and Jay quietly lined up behind them. The non-swimmer type II termers were practising the swimming motions on dry land. Dry swimming can get quite tedious after sometime and the non-swimmers began to wonder if drowning might be a better option. But there was more in store.

The II termer swimmers had been vetted by making them swim 100 meters, which was 50 meters more than they had to for passing the test. Those who were poor swimmers tired out quickly and attempted to grab the sides of the pool. These poor souls were duly rebuffed by the PT *ustads* who had been placed by the sides of the pool with poles. The poles were meant to rescue any sinker but it was used at the Academy to induce self-reliance. Accordingly, the weak swimmers were encouraged to make it back to the shallow end of the pool on their own, aided by helpful prods from the rods. They made it half drowned, gasping and wheezing for breath. Some were also vomiting.

Capt. Balhara sat there, expressionless, looking on at the proceedings. The II termers lined up at the edge of the pool, heaving and retching. Even the good swimmers were a bit out of breath. They all looked nervously at Capt. Balhara, but there was no indication on his face about what he was thinking. They should have guessed. After all Capt. Balhara was the parachute regiment which probably meant he liked jumping from heights.

The II termers were made to line up behind the five and ten meter diving board. The non-swimmers were sent to the shallow end of the pool and told to practise their strokes. Instead they huddled in a mass as if a shark attack was imminent. The PT instructors ignored them, riveted by the drama unfolding on the diving boards.

The Academy did it damnedest to remove common syndromes, complexes and fears, fear of horses, water and heights, etc. It did it the only way it knew, through terror of the alternative to not obeying commands. One could not have an officer who would show fear in front of his troops. Capt. Balhara wholeheartedly believed in this particular mission of the IMA. He also loved skydiving and could not understand why people were scared of heights.

There were some who were not scared of water, but most were scared of heights. It is but human nature to hesitate when standing on a diving board ten metres off the ground, and water. Those who were not scared of heights or water were still apprehensive of the effect water has on bare skin or testicles, if the fall was not straight. Capt. Balhara was looking for just such hesitation and he found the perfect mark to make his point.

GC Bhasin, the erstwhile BCA's pet and now just a II term toady, was an average swimmer with above average acrophobia. He had been relaxed at the prospect of swimming class. He had not expected to be told to jump off the ten meter board. He was desperately afraid of heights and had been reassured that the IMA only had swimming tests. He had surmised that the pool class would involve a little swimming, a lot of PT but no tension as such. But then the right honourable Capt. Balhara had made an appearance that had totally messed up his equanimity.

Capt. Balhara could literally smell fear; it wasn't that difficult actually. Humans and horses behaved somewhat alike when in a state of terror. They rolled their eyes, farted and fidgeted. The cadets did not quite roll their eyes; they just averted them, though some of them did feel the desperate need to defecate. Most settled for the easier option of just breaking wind.

Capt. Balhara of the 5th Battalion of the Parachute Regiment recognised all these symptoms. He had seen plenty during his time in counter-insurgency operations. He also firmly believed that an upright man should look a person in the eye, if he had nothing to hide or fear. But any normal human would hesitate to meet the eye of someone who had the power to cause you a lot of harm. Besides, the cadets did not really think Capt. Balhara operated on a human plane.

Bhasin had made a comfortable lifestyle for himself by sucking up to his superiors and volunteering for all sorts of demeaning things. He wondered if he should try that tactic now. Every ounce of common sense told him not to make any sudden movements and to definitely not look up. This was not the right time to volunteer. But then all his preservation instincts told him to do something.

The rest of the cadets just stood there with their heads down, wet and shivering because of the cool breeze and Capt. Balhara's proximity. Bhasin accidentally looked up and met Capt. Balhara's eye. He then made the mistake of doing what came naturally to him. He smiled his smarmy smile.

Capt. Balhara was mildly taken aback. Bhasin's smile would have been taken as an open invitation if he had been any other sex of any species. The direct look met Balhara's specifications, but the smile did not quite go with it. Just to sort out the visual message Balhara said, "You! Smiley! Get up to the ten metre board!"

Bhasin's smiley was wiped out. He looked around at the rest of his coursemates. Everyone looked away. Nobody smiled. Not outwardly at least.

"You waiting for a goddamned invitation card, boy?!" said Capt. Balhara as his aviators glinted dangerously.

Bhasin went up to the steps and felt a mild kick in his backside. It was Capt. Balhara's way of encouraging him.

"Would you like to hurry up, GC? There are a whole bunch of guys just eager to get up there and jump!" said Capt. Balhara. That last part may have been an exaggeration. He considered jumping from the ten metre board fun, and could not understand or tolerate people who thought otherwise.

Bhasin looked back and saw most of his comrades looking very amused now that Balhara's back was turned, though all expressions were far from eager. He somehow reached the top and looked down. That was his second mistake. He froze along with his brains and his testicles.

"GC JUMP!" said the PT instructor.

GC Bhasin's brain had disconnected itself from his body. He just stood there staring down at what looked like liquid netherworld.

"Jump, you fucking sissy!" floated Capt. Balhara's voice.

But all Bhasin could hear was the wind and all he could feel was the diving tower swaying ever so slightly with the wind. Truth be told, tall structures did sway a little, very little; too little to be noticed by anyone who was not scared shitless.

But that was nothing compared to the pitching of an aeroplane on a dark night when the rear carriage of an AN 32 plane opens up for para (chute) jumps. That being the case, it was understandable why Capt. Balhara could not empathise with Bhasin.

"If you don't jump in the next five seconds no outpass for the whole bloody month!" said Capt. Balhara in an attempt to resolve the matter through verbal counselling.

He was aware that the rest of the II termers were enjoying the spectacle and were not being gainfully employed. This he remedied by saying, "*Ustad, baki* GC rolling carry on!" Now the rest had new reason to hate Bhasin.

The ones in the pool supposedly learning to swim were sort of grateful that Bhasin was diverting everyone's attention from them. Indeed, PT instructors had lined up on both sides of the pool with long poles to goad the learners into the deep end. The same poles would be used to save them too. But it looked like the poles would not be needed today.

Capt. Balhara then devoted his full attention to Bhasin. "Jump or you will be on restrictions for the next bloody week, you idiot!"

Bhasin simply replied, "YESSIR!"

Bhasin's answer was sweet and simple, conveying his moral willingness and spiritual inability.

"I will personally come up there and kick you off, you bloody fool, if you don't fucking jump right away!" said Capt. Balhara in an attempt to provide stronger inducement.

Bhasin bleated, "Please sir..."

This simple reply might again sound ambiguous to the reader but it was clear to Balhara that Bhasin needed physical help. He had seen it many times before a para jumping exercise. Braver and tougher men had hesitated on 'green light go'[1] after looking at the swirling clouds and far away Earth. A kick at such moments in the correct dosage had been a welcome remedy. Balhara had himself administered many kicks himself and now felt that a similar moment had come. A kick was urgently required.

Capt. Balhara was moved by old memories. So moved, that he removed his aviators. The assembled audience comprising PT instructors, swimming coaches and GCs let out a muted gasp. The gasp was because Balhara removed his glasses only when he was deadly serious. He had only two moods, serious and deadly serious.

The gasp was also because of his eyes. He had close-set green eyes inherited from some Central Asian soldier who had raped and plundered his ancestors. It was also highly probable that he had inherited his serious moods from that grumpy Central Asian. Capt. Balhara's general demeanour was menacing with his aviators on and doubly creepy without them.

Looking at Capt. Balhara, the killer commando and merciless instructor, was something to be avoided. Looking into his close-set squinty eyes was quite an unnerving experience. Many a hard core insurgent had coughed up info he did not have after one look into those eyes. They say eyes are the windows to the soul. In this case, as one of Balhara's friends put it, they were windows to an asshole. Not a very pretty sight.

Those eyes were now focussing on GC Bhasin currently inching...no actually 'millimetering' on the ten metre diving board. His progress on the narrow diving board had been unsatisfactory. The diving board looked like the highway to hell and now Satan was coming up to see him off. Capt Balhara was up there right behind him in just a few bounds.

"Look at me you cowardly ass!" said Balhara as he aimed his green peepers at Bhasin.

Bhasin turned around slowly and got a double barrel blast from Balhara's gaze and took an involuntary step back. That took him one foot closer to the edge. More distance had been covered in that one second than in the last ten minutes. The crowd watching from below was impressed. Balhara advanced towards Bhasin. But Bhasin had now been inoculated. He was more terrified of the impending fall than dirty looks from scary eyes.

Balhara had expected this. He was also expected to maintain an image. In normal circumstances a DS would have counselled the jumper and invoked things like sense of honour. But in this case he rightly guessed that things like honour and pride would not have any effect on Bhasin. So he leaned over the railing and gestured to one of the swimming instructors standing below.

In a few seconds, the swimming instructor was up there, handing over the rescue rod to Capt. Balhara. The swimming instructor was a Services Level swimmer who had represented the Army. He too found the drama of sinking cadets and buoyant officer instructors highly amusing. The cadets below also found the whole spectacle somewhat amusing. Even the non-swimmers were giggling damply at Basin's plight. Truly mankind, especially the young male kind, enjoy the sight of other men's misfortune.

Bhasin took one look back and saw Satan with his pitchfork. An irritated Balhara, armed with a red-coloured rod with a hook meant for rescuing the drowning, with those creepy green eyes, truly looked like the devil in his working rig. The devil gently prodded Bhasin down the ramp. Bhasin closed his eyes, 'steeled' his will, clenched his buttocks and allowed himself to be prodded towards his doom.

Balhara was pleased with himself; he would soon show everyone how the 'Paras' handled fear of heights. Bhasin was just about three inches away from the edge and the unknown. After he had finished with Bhasin, he would concentrate on the others. There were thirty more minutes left for the class to break off and Balhara would make every second of it extremely wet.

"Come on GC move it!" said Balhara.

That effectively broke the trance and Bhasin opened his eyes. He took one look at the pool far far away and grabbed Balhara's rod. Balhara tried to pull it out of Bhasin's hands, but Bhasin held on for dear life. The drama continued for a few seconds before Balhara lost patience, let go off the rod, walked over to Bhasin, and pushed him off. Bhasin had not expected that but had quick enough reflexes to leave the rod and grab Balhara's arm. Balhara had equally quick reflexes. He grabbed the rod as it fell and warded off Bhasin's clammy clutches. But even he was not fast enough to totally avoid Bhasin's desperate talons.

Bhasin fell and as he did, he seemed to be waving some piece of cloth. It was Capt. Balhara's uniform sleeve. The cadets watching from below, enthralled as they were by the drama, did not notice Balhara's loss. Bhasin fell, still reaching for some lifeline and jack-knifed as he hit the water. The sound of his stomach slapping the water echoed off the swimming pool walls. There was a brief silence while the assemblage waited for him to surface. A nod from Balhara after a meaningful pause had the swimming instructor diving off the high board after Bhasin. Probably just to ensure that Basin did not drown too soon. It would have been in contravention of good order and military discipline.

Bhasin did surface and coughed up enough water to embarrass a whale. A tube was thrown to him at a cruel distance. It was close enough to fulfil its purpose of rescue and far enough to make Bhasin struggle, using his last few ounces of willpower to get to it. But then building willpower and mental resolve was what any Academy was about.

A Chlorine water-sodden Bhasin was fished out by the PT instructors who could barely conceal their grins. Up above on the diving board so high, Capt. Balhara was not so amused. His orders had not been instantly obeyed and the cadets were not actually displaying fear. The real devil in him was aroused along with the one that came with his looks. He squinted and focussed his green eyes on the cadets below.

The II termers were initially focussed on Bhasin's impression of a meteor. But gradually they became aware of the PT instructors stiffening up and moving purposefully towards them. The II termers outside the pool automatically started shuffling into threes[2] . Balhara spoke when he was sure he had their undivided attention. "You fuckers are scared...shit scared...unfit to become bloody officers!" After a very pregnant pause he continued, "Any fucker who hesitates to jump will be fucking relegated...now get into a fucking single file!"

The repetition of the 'F' word has many connotations in society ranging from the lascivious to the violent. As mentioned earlier, in the Army it usually means that something very serious is in the offing. It also has the salutary effect of getting males in modern society and the Academy on high alert, irrespective of the type of action impending. The PT instructors jumped to get the II termers into single file. Their efforts were largely superfluous. After all there is nothing like fear to transform a crowd into a disciplined squad. The II termers formed a single file in record time. The non-swimmers were also splashing energetically and more importantly, sincerely. Fear is truly a great motivator and promoter of honesty. It is not sure if the same would hold true in an all-woman establishment.

The II termers lined up, climbed on and jumped off, quietly. A word of advice was passed down to hold on to one's family jewels as they jumped. Jay did not take it seriously as he had jumped from the ten-metre board into the pool many times as a kid. This time however the pool seriously stung his swimming trunk's contents as they swung around in his ill-fitting swimming trunk. It was to be expected when one hit the water at 50 miles an hour. The dynamics had changed since Jay had been a kid. He realised painfully that he may have become a man.

Capt. Balhara was pleased with the performance of the rest of the cadets. He was also secretly pleased that the torn sleeve gave him a chance to show off his muscles. He left the class five minutes early. There were many who did not go to the mess for lunch that day. Burping chlorine tasting gas was not very conducive to appetite. Bhasin took a short nap wherein he dreamt he was a fish that was drowning. Quite a few of the others dreamt they were being chased by a green-eyed shark.

The non-swimmers were happy that they would live to drown another day. They had other things to worry about in the afternoon. Water was not the scariest thing in the Academy for a civilian being transformed into a soldier. They had just met a devilish instructor but they were going to meet something that was even more inhuman.

NOTES

1. 'Green light go', the green light comes on and starts blinking when the drop zone or DZ is reached and it's the signal for the jumpers to stand up, line up and hook up their static lines which help in automatically opening the parachutes by opening the ripcord.
2. 'In threes' refers to the common practice of forming up in three lines. In twos would mean in two lines.

Chapter 26

Riders in the Storm

The next afternoon they had riding classes and they would be meeting the legendary four-legged beasts of burden. These infamous beasts were the Academy horses and the cadet's burden. Jay had ridden horses as a kid and while not a very good rider, knew the basics. He was also not scared of a horse or at least getting on it, which was half the battle won. They had had a couple of perfunctory riding lessons in I term but the instructors had concentrated more on those keen on riding, leaving the others to sham it out or marvel from a distance.

Bhasin was one of those had never ridden a horse before joining the academy and had successfully avoided it in I term. After his swimming pool and the 10 metre board experience he was only worried about how tall the horse was. In fact he was still sniffling wetly as he wondered aloud if the ground really looked very far away while on top of the saddle.

The ex-NDAs knew what to expect. They had already been through the torment of riding classes at NDA. These horses had tormented and sent more cadets to the hospital than any instructor or Academy training event. Even dedicated horse lovers trod softly and rode carefully in the Academy stables. The horses were but four-legged extensions of the Army's training for suicide missions...or so the GC gossip went.

Swimming can be a nightmare for those scared of water. This category of GC would rather be screwed by a tiger than get into water of any depth. Bhasin was one of those who felt that way about both 'upper class' pursuits. There were many like Bhasin who did not believe in getting close to animals. But he was among the many who would rather try to screw a tigress than sit on a horse.

Riding was the 'done' thing in the imperial past and since times immemorial, a symbol of the upper class. It may have been a necessity designed for travelling in comfort over long distances... or at least in the times when motorised transport was not readily available. One has to admit there is something majestic about a man sitting proudly atop a horse. However there was nothing majestic about the sack of potatoes novices resembled, while bouncing desperately on top of a skittish animal. An officer was supposed to inspire awe at all times and not cut a sorry, terror-struck figure. Leadership pundits and assorted military forefathers had reasoned that such a vision would not inspire loyalty or awe among troops. The logic was weak but probably pronounced by some senior horse lover, and so riding was retained in the Academy curriculum as one of the 'should know' skills.

'Should know' was just a little ahead of 'could know' in scale of importance. That meant the skill was not really important as opposed to 'must know' and would not entail dire consequences if failed to master. It was not a test that had to be passed to graduate from the IMA. Some found it great fun but sadly, most GCs like Bhasin found the consequences of riding dire.

It could not be avoided or skipped or 'shammed out'. One just had to get through it and hopefully in one piece. The commandant of the IMA was a keen rider and had insisted the GCs learn riding before they left. The instructors suddenly started sharing their boss's passion and hence the current drive.

The GCs who did not have the aptitude for riding considered it a pain in the ass... literally. The horses, for their part probably considered the novice cadets pains in the neck, also quite literally. One buck by the horse and that was where most amateurs found themselves. The disrespect and dislike between man and animal was mutual.

The II termers gathered near the equitation officer's office. The equitation officer, Maj. Randhawa of 61 Cavalry, was not as awe-inspiring as his regiment's reputation. Almost all the officers of his regiment were international level polo riders or show jumpers. This was one of the few remaining horsed cavalry regiments in the world.

Maj. Randhawa or Randy as his brother officers knew him looked like a randy bear. He had a slight pot belly, brawny hairy arms and a heavy beard. He had none of the charm, panache or sexiness associated with the popular conception of a polo player and playboy. Historically, cavalry officers have been both, but not Randy. The GCs speculated Randy was the result of an orgy, where a bear and a horse had participated. It was likely a gorilla contributed at some stage of that party. He had a temperament that matched his physical beauty, especially since he was late for his polo match.

He did not have time to waste dealing with the GCs and so he told the Riding JCO Risaldar[1] Noor Mohammed, "*In* GCs *ko accha* riding *karao, pasina aana chahiye-dekhna kisi ko ziyada chot na aaye*!"[2]

Risaldar Noor Mohammed heard and obeyed. He truly loved horses and thought them better mammals than most humans, especially cadets. The cadets were prone to treat his beloved horses like pack animals... and if it were up to him, he would have used the cadets to groom the horses, not ride them. There was nothing much he could do about it apart from making sure the cadets had a hard, short time on the horses. The secondary aim being the hope that neither got injured-at least not the horses. The end result, however, was that both horse and cadet were discouraged and disenchanted by the entire idea of riding.

The horses at any training institution have been ridden by generations of novices and they instinctively knew the calibre of the man on top of them. They also knew exactly which way their stalls in the stables were, and the shortest way thereof. If any water troughs or fences came along their way back to their stalls that was just too bad. In any case the horses knew their burdens would have lightened by the time they reached.

The riding staff was mainly from the horsed cavalry, RVC (Veterinary Corps), the ASC (Army Service Corps or the supply and transport guys) and occasionally from the PBG (President's Body Guard). These types of corps were either ceremonial, like the 61st cavalry and the PBG or relatively cushy and lucrative like the ASC.

These glamorous polo-playing peacetime soldiers were also generally looked down upon, envied or ignored in turns. They were not considered real soldiers by the other fighting arms. The riding classes were the only time instructors from these corps got any importance. It was one of the few times they could wield real power and be treated with some deference.

Swimming and/or jumping off the high board made some feel like pissing in one's pants or trunks, as the case may be. Riding on the other hand was a different kind of terror. Imagine a five-hundred kilo beast that had a mind of its own. It was comparable to riding a 500cc motorbike that had a tendency to do what it wanted, like throwing you off, running you down or maybe even biting you...the thought often made some of the cadets feel like crapping in their Jodhpurs.

These thoughts were uppermost in the cadets minds as they stood there listening to Rissaldar Noor Mohammed. The good JCO was rattling out his routine list of the many don'ts and a few dos. The animals had no such restrictions. The highlights of the introductory speech were the points of the horse-the finest animal...make that creature, according to him, in the world. The II termers were not impressed, and they were shuffling around adjusting their *puttees* and helmets. A few like Bhasin sought last minute whispered reassurances from his neighbour, who was busy praying. He kept reciting the Hanuman Chalisa, till he was shut up by an NCO.

Rissaldar Noor Mohammed was incensed. These cadets were not at all interested in riding. They just wanted to waste his and the 'poor' horses' time. A few of the good riders had already presented their credentials to the riding NCOs and had been given appropriate horses. The mount allotted would depend on what they wanted to practise, jumping, polo or tent pegging. The rest would to be sorted out appropriately.

The NCOs formed groups based on riding proficiency. It was basically three groups—the novices, those who had sat on a horse earlier and actually ridden and the expert riders. The last category was generally familiar to the riding instructors as the military riding fraternity was not very large. They were therefore given decent horses and left to their own devices. A few opted to join the polo game in progress in the adjoining field.

The remaining cadets stood in their respective groups. They were standing in groups in threes, in an orderly fashion. The novices' group still managed to look like a terrified huddle in spite of its military formation. They looked on nervously as the horses approached. The sight of the huge creatures being led in by the grooms did not excite them as it did the expert riders group.

The champion riders cantered around confidently watching their coursemates' discomfiture. They actually loved riding and could not really understand or remember the phobia of novices. The animals swished their tails, stamped their huge steel-shod hooves and bared their massive teeth. It made the novices wonder if there was some way to outrun a fate that entailed getting run over by a horse. Most cadets at the IMA had had some versions of such morbid thoughts in the last 60 years or so.

GC Kapoor was one of the few who were determined to learn riding. He actually listened to the JCO's instructions. GC Bhasin decided to follow Kapoor's example. His intentions were more simplistic. He just wanted to get on the Rissaldar's good side so he would get a better time or at least a safer one. His sincerity was noticed by Risaldar Noor Mohammed and he nodded in appreciation. There are many ways to suck up to those who matter. Looking like you are hanging on to their every word was one of them.

"*Sahib, accha ghoda dena*[3]..." simpered Bhasin.

Rissaldar Noor Mohammed was disappointed by Bhasin and said curtly, "*Sare ghode acche hain sahib, siraf rider kharab hota hain*[4]!"

Then seeing Bhasin's worried look continued, "*Aap ko bhi accha ghoda hi denge hum*[5]!"

Bhasin was not at all convinced that all horses were good. The cadets moved to their horses cautiously as if they were rabid. And then Risaldar Noor impatiently bawled out, "GC mount!"

Most of the GCs had trouble mounting. The stirrups were deliberately kept high by the *syces*[6] who also wanted their pound of entertainment. It is very difficult trying to put one leg in the stirrup while stretching your foot as high as it could possibly go. To add to the difficulty, the horse kept fidgeting and moving.

One of the cadet's feet got stuck in the stirrup and he lost his balance. The horse moved away while he was trying to mount and he lost his grip on the saddle and the reins. He was rescued almost immediately...well almost immediately by the NCO and grooms. They had been expecting just such an occurrence and waited till the cadet had been dragged around a little and terrified a lot.

Everyone finally got to mount and they sat uneasily on their horses. They had all heard the horror stories and legends. Advice had come from all directions, especially from the Academy riding champions watching from the sidelines and enjoying themselves immensely. It was rumoured that the position of the horse's ears was an indication of the horse's intentions. But then so was the tail swishing and eyeball rolling, all things horses did anyway. So it was but natural that the cadets were concentrating more on looking for tell-tale signs of the horse's intentions than listen to the riding instructors. The riding NCOs knew this but they were not hurt by this lack of attention. They had experienced this with every course and had also perfected their revenge.

"Horses walk!" the NCOs yelled once everyone had mounted, almost everyone that is.

The horses obeyed and the riders scrambled to settle down in the saddles. It wasn't so bad actually... at least that was the general consensus in the initial few minutes. Sitting on a horse that was just walking did not seem so bad...sort of like sitting on an undulating sofa. After a little dust had been raised and the cadets had actually started getting used to the idea of sitting on a horse, the next order came.

"Horses trot!" said the riding NCO and the misery came...

For the uninitiated, the trot can be quite an agony. A horse has four legs and a car, four wheels, as everyone knows. But a lot of people make the mistake of assuming the horse is a mode of transport like a car and thus a similar ride. But unlike a car, where all four wheels roll at the same time, the horse's legs move almost independently of each other. And to make matters worse not all its hooves are on the ground at the same time. A trotting horse looks very nice and elegant with its hooves tripping up rhythmically. For that image to hold true the rider has to sit correctly and not be a sweat-streaked, terrified sack of bouncing meat.

Expecting someone to sit correctly on a hard slippery piece of leather called the saddle, while bouncing around uncontrollably, is downright unrealistic. The correct way was to move up and down with the rhythm of the horse. The instructors had mentioned it, but the cadets had either not paid attention or had not understood. The only mammal most of the cadets had sat astride on was probably their father many, many years ago. A fond parent trying to bond is very different from a malicious mammal that wants to get rid of you.

A few of the riding *ustads* tried to explain and even demonstrate how to sit correctly on the saddle, but it all fell on deaf ears. It is actually very difficult to hear or see much when bouncing on a hard object that is fairly high off the ground. Some of the cadets tried to move up and down as demonstrated by the instructor. But they just could not get into the rhythm because their main focus was just staying on the horse, not riding the damn animal.

To put that in perspective, it is hard to explain the act of making love to someone who has never held a girl's hand. Basically it meant they got it all wrong. But the instructors weren't too cut up, since it added to their day's fun quotient. There was precious little fun a riding instructor could get in the Academy. They yelled instructions that no one understood-they may as well have been neighing.

GC Kapoor was one of those who were trying very sincerely. He was getting the hang of gripping the horse with his thighs while using the stirrups to go up and down with the rhythm. But sadly in the process of concentrating on his legs, his grip on the reins loosened. It was for just a few seconds, but it was enough for the quirky quadruped to seize the opportunity. It gave one victorious neigh and took off. And it took off straight for the stables with a desperate Kapoor on its back.

Kapoor tried very hard to remember what the instructor had said and gripped the horse's sides, riding it like Harry Potter on his broom. The horse had been at the Academy for the last six years or more, had seen many cadets and been ridden by better riders than poor Kapoor. It decided to jump over the water trough which was in the way. That move was guaranteed to lose the stickiest rider. Kapoor realised he was heading for a wet landing and if he missed that, a very hard and painful, possibly dangerous one. He racked his brain, closed his eyes, pulled back hard on the reins and leaned back. The horse slowed down and a groom caught it. It was just in time as Kapoor had already lost a stirrup and was just about to slide off.

A wobbly-legged Kapoor got off and sat down on the nearest bale of hay. The rest of the horses had gotten spooked and were in various states of unrest. Bhasin's horse reared up once and landed heavily on its forelegs. That effectively unseated Bhasin from the saddle and he was left stranded on the horse's neck. The horse then decided to make a dash for the stables with Bhasin screaming for help.

A few minutes later a shaken up Bhasin was led back by one of the NCOs. As Bhasin passed Risaldar Noor Mohammed, he gave him a hurt look and said, "*Sahib dhoka diya hai aapne...wo bahut kharab ghoda tha*[7]!"

Risaldar Noor took Bhasin's heartfelt critique with a smile that could easily be interpreted as 'mission accomplished'. It was a testimonial to his efforts to back the horses in the war against humanity and cadet kind. The horses had been winning so far in all the officer training Academies.

Jay was part of the group that had some proficiency in riding. They were having a relatively easier time, being able to control and sit on their mounts better. They felt a sense of superiority and relief, which they expressed by way of giggles and smirks. They had all heard Bhasin's parting shot and were very amused.

Just then Capt. Ali Shah cantered up. He had been playing polo and had just changed his pony during a break. He had noticed the happenings and had decided to take a look. He had not liked what he saw. Unfortunately he had also heard the disparaging remarks and jokes about Bhasin.

Capt. Shah was a great believer in course spirit. He was deeply offended by the apparent lack of sympathy for a wounded comrade. He cantered around the group swinging his polo mallet languidly. Jay and the rest of the II termers were fascinated and terrified. They were fascinated by the impressive sight of the good-looking polo player. Capt. Shah did cut a dashing figure. Jay and the others wanted to become like Capt. Ali Shah. But they definitely did not want to become his victims. They were worried because he was riding around the group without any apparent purpose. It's always a bit scary having a shark circling you.

Some of the 'expert riders' who had hung around to watch the fun had the horse sense to smartly disappear from human and equine sight. That left only the so-called riders group which Jay was a part of, stranded on their horses. Even the horses sensed the gravity of the situation and stopped fidgeting. They probably realised their normal options of running away would not be without reprisal and kept their heads down...literally.

"Noor *Sahib yeh* GC *aur ghode dono ko masti aa rahi hai...*" said Capt. Ali Shah, "*inka thoda pasina nikalo*[8]!"

Risaldar Noor Mohammed was happy. He now had the authority and the fear of an officer instructor to back up his orders. He looked at the other NCO instructors and saw that they were in accord. He gestured to one of the other NCOs to take over the beginners and went over to Jay's group and said, "GC *rakab* up! (Stirrups up)

A Botswanian cadet next to Jay looked unsure what to do next. GC Aurora, who was next to the nervous Nubian helpfully suggested, "He means lift up your reins."

The Botswanian promptly did as advised and was next seen heading for the stables at high speed screaming, "*Rakab* not working! *Rakab* not working!" Aurora smirked with the deep satisfaction of a successful sadist.

The rest of the group put their stirrups up. They were quite confident of their riding prowess and thought they could manage a little bit of riding without the stirrups. The next order was not long in coming.

"TROT!" bawled Risaldar Noor suddenly.

The horses trotted obediently. After a few minutes of trotting, Jay and the rest of his band understood the beauty of Risaldar Noor Mohammed's plan. The bumping without the support of stirrups was mind-numbing as also tear and ball jerking. That, combined with the continuous effort to stay on the horse without sliding off, had them all sweating from every hole and pore. After five minutes the group was ready to surrender...whatever would help stop this brain-rattling locomotion.

Bhasin and the rest of the beginners felt vindicated and relieved. Their turn would come but not today. The class got over soon after. Jay himself could not have said when. Time for him had come to a standstill. All he was focussed on was the large animal between his thighs that seemed bent on breaking up his body one bump at a time. The horses however knew the signal for the end of training and they were ready.

Most of the cadets had concentrated on staying on the horse rather than hold their reins securely. Not that it would have mattered much to the horses either way. But it did matter to most of the cadets the moment the whistle blew. Almost all the horses tried to head back for the stables. Some cadets managed to control their horses; some let the horses have their way and hoped they would not fall in a pile of horse dung.

Jay's group had a doubly tough time as their stirrups were up. There was utter chaos, dust and filthy abuses as the cadets tried to get their feet inside the stirrups. It is unlikely that horses had ever been called by so many different names that had some sexual or incestuous inference.

The NCOs shouted at them to pull the reins to a side to control the horses, but it was too late for most riders. The theory was if you turned the horse's head it would not be able to run away. The theory was good, but the horses were, for the most part, stronger and generally did not believe in theory. Still Jay and most of the cadets were able to get back without any serious injuries.

It was another hectic five minutes before they all managed to dismount, give the report and put some distance between them and the dastardly stables. As Jay and Kapoor shuffled on wobbly legs back to their cabins there was only one thing on their minds.

After they reached the platoon quadrangle and were trying to take off their *puttees*, Jay broke the silence, "Do you think we will be able to reproduce after all that bumping on those bloody animals?"

"Fuck the reproduction...I hope I'm able to sit within the next few days...even my asshole is numb!" replied Kapoor somewhat shakily.

The next day Aurora walked past them legs slightly apart. "Don't even ask!" he snarled at them. Jay didn't ask but was dying to discuss it. After all the sight of the normally supercilious Aurora unable to strut and walking in a 'remarkable' manner was a strong discussion point.

"What the hell is his problem?!" asked Jay.

"Folliculitis also known colloquially as '*bal tod*'!" was the simple reply. It was usually caused by poor hygiene and the heavy-duty friction of horse-riding. The horses did have the last word.

NOTES

1. Risaldar is the same as Subedar and is used by the cavalry and the armoured corps for their JCO ranks.
2. Literally 'Make these GCs ride hard and sweat and make sure nobody gets hurt!'
3. Sir give me a decent horse.
4. All horses are good sir, only riders are bad.
5. We will give you also a good horse.
6. Syces—is an Indian variant of the term groom, more or less the same functions though in the army they are used to denote a step up from groom often used to train or break in horses.
7. Sahib you have betrayed me...that was a very bad horse.
8. Noor Sahib the horses and GCs are both feeling frisky...make em sweat a bit!

Chapter 27

INTO THIN AIR

Chib was reading a book very seriously. Unfortunately he was reading in the auditorium, which was a serious matter. The entire Academy was there for a 'guest' lecture and Jay and Kapoor, sitting next to Chib, were justifiably nervous. Guest lectures in the central auditorium were a good time to relax, even catch up on your sleep. If you were the sort who slept like a cat with inbuilt radar, you could sleep with one or both eyes open.

One was expected to look attentive or at least pretend to. The only other fairly simple requirement was to maintain absolute silence. Those afflicted with a cough had a rough time trying to control it. One was expected to exercise self control or the 'control' would be taken over. The cougher would be taken outside for a round of pushups, etc. It worked on the theory that if you are breathing hard from exertions, it left you too winded to cough.

The general idea was to give the guest speaker the feeling that he had the undivided attention of his cadet audience. These speakers were usually senior officers or war heroes of some sort and generally touchy. It was vital that these guests did not get offended by the lack of audience or attention. Both were therefore forced.

It was a different issue that almost everyone, including the officer instructors, tried to catch a few winks during the lecture. The fact that the stage was lit up while the rest of the auditorium was dark was conducive to these short snoozes. In any case, the speakers had also been through the mill themselves and knew that the audience would be sleeping. The larger hearted did not mind. But those with bigger egos and smaller minds did. All senior speakers did mind having an unappreciative audience.

Jay and Kapoor however had reason to be concerned. They were in the fourth row where the lighting was just enough to read. But that also meant they were just two rows behind the Academy appointments and the commandant. It was not unknown for the occasional bored instructor to prowl the aisles. It was a good way to catch inattentive offenders such as Chib...and kill sleep.

"What the hell are you bloody reading?" asked Jay.

"It's a book by Jon Krakauer or somebody... called "Into Thin Air'...you mind?!" asked Chib cheekily.

"You silly SOB, if some appointment or DS sees you he'll have all of us bloody rolling throughout the damned lecture!" said Kapoor.

"Is it so fucking important that you have to read saala military history assignments here!" asked Jay.

"It's not military history you ignorant pussies, it's about the fucking Mount bloody Everest and a tragic climbing expedition!" replied Chib with a superior look.

"What's up? You're not feeling well? I thought you read only pornographic material, so your brain doesn't get too taxed by unfamiliar story lines!" commented Kapoor sarcastically.

"Why the hell are you reading about Mount Everest when it's not a tactical exercise, you dumb shit?" asked Jay of Chib once more.

The conversation was nipped just as it was getting more interesting. The ACA (academy cadet adjutant) turned around and looked directly at them. ACA Ranjeet had blue eyes and a bland expression. He was known to be unmoved by human suffering. He was Capt. Balhara's distant cousin hence the ocular malevolence was genetic. The three 'shut the fuck up' immediately as visually commanded by the ACA. They also reverently hoped he would not remember them later.

The ACA was the most powerful cadet appointment in the Academy. He could do a lot since he had a direct line to the Academy adjutant[1] . In this jungle he could cross battalion boundaries in search of his prey. The prey Kapoor, Jay and Chib shrank into their seats and slept fitfully with one eye open, literally.

After the lecture, on the way to the mess for dinner, Jay punched Chib in the arm and asked, "Why the fucks did you have to read that damned book anyway?"

"You mean you *chutiyas* don't know!" answered Chib in a self-satisfied tone. "We are due for a two day, three night mountain exercise *'Parvat Raksha'* (mountain defence) next week."

Jay and Kapoor looked at each other and felt the same sensation. The tingling sensation commandos and other such heroes get when they are finally told of a daredevil mission. This was one of the rare times they would actually be doing some soldiering. Soldiering...what they had joined the Academy for in the first place. Had they done some analysis, they would have realised that there would be very less daring and more of devils involved in an Academy training exercise. They entered the mess and found the rest of the II termers discussing the upcoming exercise. There was some excitement over actually learning and practising tactics in the mountains.

The ex NDA II termers added their two bits to the discussion. Aurora was leading the discussion pooh-poohing the upcoming exercise. The NDA had a camp called 'Rover' which was supposed to be the toughest military training camp in the world for teenage types. Some cadets had actually died during Rovers in the dim past and thus any other camp was passé and sissy.

A minor scuffle broke out when a DE remarked that 'Rover' sounded like some obedience camp for dogs and maybe that was why it was intended for NDA cadets. Things began to get tense and the III termers had to intervene. But they too got into a heated argument about the toughness of Camp Rover. The DEs had the last word when someone said Rover was but a camp for cadets while 'Parvat Raksha' was an exercise for gentlemen cadets. There would have been more violent disagreements had the duty officer not made his timely yet unwanted appearance. That however did not save the I termers who did not give a damn either way.

The BCA, an ex-NDA, had seen some of them sniggering at their senior's shenanigans. He came to the somewhat correct assumption that they were mocking the childishness of the issue. He promptly ordered them outside for a bracing round of pushups to let them know that Camp Rover was too sacred an issue to be mocked. He also did it to demonstrate to the duty officer his command and control over a possibly unruly situation with unofficerlike behaviour.

The BUO, a DE, also took affront. He was upset on two counts. Firstly, the BCA, a senior cadet appointment was lowering himself by taking sides in the eternal DE and ex-NDA rivalry. And secondly and most importantly, the BCA had managed to impress the duty officer with his masterful handling of cadet discipline. He had thus 'stolen a march'[2], so to speak. The BUO seethed with the repressed feelings of thwarted ambition. He remedied the situation by ordering all the II termers outside to join their I termer brethren in pushup land.

The duty officer realised what was up. He was from OTA, a short service officer who had now gotten permanent commission. He loathed the superior attitudes of most of the ex-NDAs and some of the DEs. He also strongly disapproved of cadets behaving like rowdies. But he disapproved even more of appointments trying to impress officers by screwing their juniors. He ordered all the III termers to report back in FSMO[3], in ten minutes and no exceptions.

That meant all appointments, except the BUO, BCA, JUOs and the CSM, would be getting their undies in a bunch. In this case it also meant their understudies. The III termers blamed the BCA and BUO, correctly, for their suffering and vowed there would be consequences. The scope of the original discussion of which type of entry, DE or ex-NDA, was superior had just been widened several fold.

There was a mad rush as the III termers scrambled to get to their cabins and get ready. At that stage nobody gave a damn whether the NDA camps were superior or the USA's. The other terms also just wanted to get back to their cabins and beds. Everyone heartily regretted ever having heard the name Rover or Parvat Raksha or even Guantanamo.

The next day the subdued II termers fell in, in front of the CSM's room to be briefed. The CSM had a list of camp appointments. They were to be given temporary appointments as section commanders and the like. The section commander's main job, in this case, would be to designate who would carry the LMG (light machine gun) grenade launcher, maps and other control stores[4]. It was a position of command, albeit limited command, but Aurora did not realise that. Most likely he did not want to.

Aurora took his job of section commander, at least the 'commander' part, seriously. It's possible he considered himself the rightful heir to Napoleon's legacy. Unfortunately, he decided to do this by following Hitler's methods. The other II termers were not at all interested in Aurora's ambitions or command potential. They were just excited by the prospect of some military training. It felt like they were going on some campaign. The III termers saw the enthusiasm and kept their silence. They had been there, done that and some had been done for.

Aurora found that his new-found command wasn't as empowering as he had hoped. One of the section commander's jobs was collection of all the stores required and its distribution. Aurora reported to the III termer who was the CQMS (Company Quarter Master Sergeant) to collect the stores. He knocked at his door and waited till the CQMS told him to come in.

"Sir I've come for the camp stores..." said Aurora diffidently.

"There...pick up all that shit and get it out of my room pronto!" said the CQMS pointing at the rolls of maps, rolled up bivouacs, poles, buckets of blank rounds, compasses and other such camp-like things.

He then noticed Aurora was alone. "You...planning to carry all this crap by yourself?!"

Aurora saw it as an opportunity to whine and crib about his recalcitrant command. "Sir, I'm alone...there's nobody..."

But the CQMS was not interested, "Get the fuck out and get somebody, I don't give a damn!" He then continued, "Get all this shit out of my room right now or you can explain to the DS why you did not collect the stores. You got bloody ten minutes... after that I'm going for dinner! Got it?"He also added 'asshole' in an audible undertone.

Aurora got it, but still could not come to terms with the unfairness of the whole situation. He lugged all the stuff out of the room himself. The CQMS threw the maps at his face and told him to count everything and sign the receipt. He then shut the door abruptly while Aurora stood there feeling very aggrieved and peeved. Aurora was positive that was not how Napoleon had been treated.

Aurora looked around and spotted a I termer scurrying around in the shadows. He yelled out just as he tried to melt into the shadows. Unfortunately for the, I termer he was collared before he could make it. Aurora used him to ferry some of the stuff back to his room before the I termer slunk off. Fortunately for him he was not wearing a name tab, and knew there was no way Aurora could track him down, soon.

Aurora, the great section commander, looked around for his command, someone...anyone. He saw Sandy looking at him languidly from the doorway of his cabin. He was eating a banana suggestively and looked stupid. There were very few people who could look languid and stupid while giving suggestive looks. However the only thing Aurora understood from Sandy's look was that this was an idiot who could be manipulated.

Sandy said, "What's that stuff?" and then answered his rhetorical question, "The camp stores, *hai na*!?"

Aurora choked off all the nasty retorts he had and smiled. "Yeah! You know I'm the section commander...and maybe you could..."

Sandy was a dumbass, and everyone except his mother agreed on that count. But even he had a keenly developed sense of survival. He quickly realised that Aurora was now a friend in need who had climbed the first rung of leadership. He also concluded somewhat correctly that a meaningful relationship with Aurora was fraught with useful possibilities. But these things came at a price, and the price for most things in the Academy was measured in sweat.

"So?" said Sandy cautiously.

Aurora decided to lay his cards on the table, "Listen you help me and I will make you my..." He was a bit lost. There was a limit to what he could do, while commanding only nine men, to gratify faithful followers. There was not much to offer by way of pelf and privilege.

Sandy vaguely understood Aurora's plight but was very clear on what he wanted. "No trench digging for me."

Aurora smiled the smile of a man who has managed to con a prostitute, and said "Sure man anything. Just help me with this stuff and we'll see."

The day of the exercise arrived. The cadets had all fallen in. Aurora and his mighty section had fallen behind-in many ways. Aurora had collected the breakfast packets for the section and had found one short while distributing it. Chib had filched one from Aurora's pack out of spite. There was much recrimination and as section leader he was told he ought to go hungry before his command. A valuable lesson was learnt though the wrong point was taken. Aurora vowed revenge and dark retribution. In the melee, he forgot to issue the blank rounds and other stores till the instructor appeared.

The appearance of the instructors sent all the wannabe generals and commanders of all levels into a tizzy. The section commanders scrambled to confirm that all the control stores were there. Aurora distributed the maps, compasses, blank rounds and blank firing attachments with the help of his newfound henchman, Sandy. It was only after distribution that he realised that sidekick Sandy had not written down who had been given what. Without that vital record, getting back all the stores would be next to impossible.

There was no such thing as gentlemanliness or humanity when it came to accounting for control stores. Aurora seethed and gritted his teeth, much as

Sandy's father and sisters had in the past. Sandy, as always, had set a new benchmark for sheer incompetence. Aurora seethed because he was unable to do much, due to internationally recognised standards of use of retributive violence in civil society. Also Aurora realised it was as much his fault as Sandy's.

The instructor came around and asked Aurora if everything was okay. Aurora stiffened up, puffed out his chest and said, "All under control sir!"

Sandy in his new-found role as deputy general minion echoed him, "Yes sir, all under control!"

"Good josh...carry on!" said the instructor to Sandy and moved on to the next section.

Aurora was very, in fact terribly, pissed at Sandy for trying to steal his thunder. Jay and Kapoor who heard the entire exchange could not help but giggle disrespectfully. Actually they just giggled and Aurora found it disrespectful. As section commander he had one last weapon in his armoury, both literally and figuratively, to sort out disrespectful underlings.

The LMG had to be carried by someone and it weighed two-and-a-half times as much as a rifle, and had a dozen magazines to boot. Most tried to shirk the responsibility and the usual method was to take turns in sequence. But human nature being what it was, it was very difficult to locate the person who was supposed to take over the burden from you. Aurora knew that and deputed Jay to carry the LMG and Kapoor to take over from him, but did not elaborate who came after that.

Jay asked Aurora politely, as befitting a leader of his stature, "And who the hell is going to take over the LMG after that?"

Aurora mumbled a reply and rushed off to give a report to the II termer who was doing the job of platoon commander. Jay and Kapoor looked at each other and then at Sandy who was looking extremely smug and self-satisfied.

"What's your deal man, you look like the cat that's got the cream?" asked Kapoor.

Sandy laughed knowingly. And the rest of the section realised here was the first of their course who had sold his soul for personal gain.

Kapoor said, "In fact you look positively like someone who has been licking a lot of things apart from cream."

They heard distant screaming, and realised that the new regime consisting of first-time leaders was having a tough time. The II termers fell in neatly; they were well disciplined and somewhat trained soldiers after all. There were a few exceptions. A couple of them were caught posing and taking pictures of themselves with their weapons and fierce looks.

These pictures were meant to impress girlfriends, sisters, mothers and other such impressionable females back home. The instructors did not find it or their motivations cute or funny. This was especially aggravating since the antics were holding up the report. The photographers and their subjects were now doing pushups with the very weapons they had posed with.

The fall-in and reporting process finished off without too much by way of sweat, tears and frustration. And then the order came to mount up. Spirits were high as the II termers clambered onto the trucks. There were some of the usual 'last minute' tense moments as the section commanders scrambled to check if all their 'men' and. more importantly, material, were present and accounted for. There was more scrambling as Sandy realised he had forgotten his bayonet in his cabin.

Aurora went around trying to ensure that all the 'control items' were still under control. Not an easy task considering his Army of ten was scattered in three trucks. Aurora was justifiably concerned; after all, the buck stopped with him as a leader. This was the first thing one learnt in the Academy about being an officer. Victory was important, but covering your ass was very, very important. See what happened to the losers of the World Wars.

The II termers sang martial songs from old Hindi movies. They did feel like some conquering Army. These feelings were not shared by the DSs following the convoy in a jeep. They were particularly incensed by the cadets singing louder as they passed a girls college. The cadets tossing breakfast wrappers out of the truck did not help either. The unanimous thought among the officers was 'Bloody unofficer like behaviour'.

The DSs were in a mood to mete out '*ragda*' when the convoy finally reached. The moment they alighted they were met by the HOTT (the head of training team). This was a crusty old brigadier. His first words were, "Did you hear the boys singing...great josh! Good to see such high spirits before a battle!"

The HOTT's words changed the mood among the DSs instantly. They all chorused in agreement "Yessir good josh among the cadets sir" and other faux words to that effect. Getting one's mind in sync with one's superior is something learnt very fast in any military. The DSs abandoned their plan to sort out the GCs right then-they had thirty six hours for that.

The cadets dismounted from the trucks unaware of their narrow escape. There was the usual confusion found among semi-trained soldiers. In Jay's section however, apart from the confusion, there was also animosity. This was in large part because of Sandy's attitude. Sandy had taken Aurora's vague election promises very seriously. He was now refusing to help the section in any way. Soon after the report, Sandy just upped and vanished leaving the rest of the section to lug all the stores up the mountain.

Aurora was faced with a mutiny and that too on the very first day of his command. He was ready to kill Sandy himself. He was arguing with Chib when Capt. Shah walked up and asked what happened.

Aurora took the lead and decided to cover his ass. "Sir the section... I was telling the section to carry the loads..."

Capt. Shah realised Aurora was expecting him to step in and shore up the sorry section commander's authority. He also realised there was a marked absence of team spirit. He was disappointed by Aurora's lack of command potential. Any leader who expects his senior to back him up, has already lost the moral authority to command. Aurora and his section were new in the Army, but already sensed this leadership axiom.

"Is your entire section here...I see you're one chap short." observed Capt. Shah, rhetorically.

Chib piped up, "Sir GC Sandeep isn't here!" The others looked at him, mildly disapproving, as it was strictly incorrect form to give away one's team member. But this was Sandy so no one truly minded.

Capt. Shah made up his mind after a few tense seconds. Seconds during which Aurora realised he had made a career-killing mistake by involving Shah.

"Sir, we will manage..." said Aurora as he attempted damage control.

Capt. Shah went up to a boulder that must have weighed at least 15 kilos. The II termers watching cringed inwardly. They knew exactly what Capt. Shah had in mind. The other II termers quickly loaded up their stores and started walking up the mountain, without any eye contact. Aurora and his section waited for the sword to fall. "OK! Aurora, ensure you and your section gets this stone...along with your stores up the mountain." And just when they thought he had finished, Capt. Shah added, "You have three hours!"

And so began their long slog up the mountain. Jay had the first shift. He felt his eyeballs were going to pop out of their sockets because of the strain. No one spoke, everyone was breathing too hard to think of talking. Other sections that passed them did not say a word. In normal circumstances there would have been a ribald comment or comparisons to a female or male organ.

The other II termers knew, but for the grace of God, it could have been any of them in a similar situation. Some of them offered to help carry the weapons and digging tools. The offers were made even though if caught the DS would screw both the helper and the helped; but even so these offers were given and gratefully accepted, inspite of the risks. After all camaraderie had been dinned into them and was now very strong.

Aurora took over the boulder from Jay when he stumbled. After a minutes rest, they started off again.

"You're a *chutiya* you know that." said Jay to Aurora.

"And you are a *gandu*, you know," replied Aurora.

"Yeah and now *hum sab ki gand le li saale* Shah *ne*[5]!" said Kapoor.

Everyone started laughing and somehow things seemed better. The weather was beautiful and they were young and carefree...if not stone free. They had a quick break while everyone posed with the stone and photographs were taken. These pictures were ostensibly to impress parents and other such sentimental well-wishers. Civilian relatives, especially the maternal sort, were sometimes gullible enough to believe carrying big boulders up mountains was an essential part of military training and associated heroics.

The burdened section managed to reach in four hours. They felt somewhat stoned and groggy. They hoped Shah would not be there waiting for them. But Capt. Shah was indeed waiting for them. Sandy was also there and he was carrying four pickaxes and five shovels along with a water tank on his head. He was sweating and pouting, a combination only Sandy could manage.

Capt. Shah said, "I hope you chaps have learnt something today?"

The section, minus Sandy, chorused "Yes sir" loudly. They were tired, drenched in sweat, cold, thirsty and hungry. But they were in high spirits and adrenaline charged, ready for anything. Capt. Shah recognised the mood and smiled, this was the kind of spirit he expected in gentlemen and cadets.

"All right, fuck off to your section post...and make sure the defences and section commander's orders are ready tomorrow." With that, Capt Shah walked off. He said something to Sandy as he passed him.

The moment Capt. Shah disappeared, Sandy collapsed. The rest of them picked up the stores and Aurora recovered some of his authority long enough to start passing instructions. Everyone was too tired to argue. It was sometimes easier to just follow orders and let the leader do the thinking.

Aurora kicked Sandy and asked him what Shah had said. Sandy was only too happy to inform Aurora, "He said he would listen to section commander's orders from everyone."

Aurora was a bit peeved that he would not be the prima donna. The others were also peeved; now they would have to mug up the landmarks and orders. Uneasy lies the head that wears the crown- or section commander's armband, as it was now. In this case, all the heads would lie uneasy because, come morning, all of them would have to wear the prickly crown in turn.

The digging commenced at feverish pace. Luckily they found the last camp's half-filled trenches, so the job was relatively easy. Sandy kept moaning about how much he had suffered... nobody had the energy to physically shut him up.

NOTES

1. The ACA or Academy cadet adjutant literally had a direct line since he was the only appointment with a telephone in his cabin. He also spoke directly to the IMA adjutant when the occasion demanded—which was quite often.
2. In the final ie third term the competition among the cadet appointments was intense as the final merit would be decided not only by written or other marks but also by marks allotted by the instructor, which could make the difference in who would come into the 'superblock;' or the first ten
3. FSMO (field service marching order) comprises of a small pack and magazine belt. It entails getting to the room, changing from mess dress into field dress, assembling the pack, wearing it and getting back—all this in 5 minutes, but can can be done with practice and prior preparation.
4. 'Control item' was a generic term used for items that were normally kept in a 'kote' and issued when required. Apart from weapons, it also included items like compasses, bayonets, grenade launchers, blank firing attachments and even pointer staffs (called staff, but actually a folding metal frame used to point targets to another person).
5. A rough translation would be 'and now that bloody Shah had screwed us all in the ass'.

Chapter 28

A SOLDIER'S LIFE

At night, on top of a mountain, Jay and Kapoor sat and watched the star-studded sky as they stood the first shift of guard duty. The section had been paired into 'buddy pairs'. It was standard army practice in the field to ensure accountability and safety. There both had the same thought 'If only my buddy were a woman how romantic this moment would be'. But they settled for hot tea instead of romance and feminine company. The human body, if not the mind, is more realistic in prioritising needs.

The mood changed along with the weather, as it often does in the mountains. There was a thunderstorm and within seconds, the section's defences were swirling in torrents of water. The II termers had fortunately brought their rain capes, thanks to the sage advice of the third termers. But to cut down weight they had not felt the need to get extra clothing and socks and were sorely regretting it now. The rain cape did its job for about twenty minutes and then gave up.

Aurora was trying his best to save all the maps and other 'control stores' from the rain. The life of a supreme commander is not always easy. Jay and Kapoor tried to help Aurora by holding all the weapon barrels up under the rain capes. They spent the next four hours standing like that, till the rains finally abated. In an actual operation they would have reacted in a similar manner, giving priority to saving their weapons and 'control stores' hardware. That would have been because the hardware would save their lives. But now they were doing it to save their ass. It was a good training technique to understand priorities.

They sat or reclined and tried to sleep as best as they could on the wet muddy ground in their damp rain capes. Then the wind started blowing and froze them stiff. Sleep was not possible; it was something of a distant dream. The II termers drifted in and out of a comatose dreamless state. And then came dawn and the sun rose.

The II termers gathered stiffly next to the '*chai* thermos' which had been brought up by the orderlies. The hot tea made them feel some of their more distant extremities again. Sandy was despatched with some extra tea for those still guarding the section defences, and their precious control stores. Jay and Kapoor collected everyone's packed breakfast, which was cold omelettes, crumbly stale bread mashed with butter and jam and some pickle. But they ate the mess, ate it all and ate it gratefully.

The combination of food, a little sunlight and food made them feel better about life again. They sat next to the trenches and dozed off. Nobody remembered Capt. Shah or his section briefing. But Capt. Shah did...

Aurora slept fitfully and then he suddenly woke and jumped to his feet. "Oh fuck, fuck fuck fuck!" he kept saying.

"Listen man stop sounding like a panicked chicken clucking fuck...what's up?" asked Jay.

"Shah Sahib is gonna be here any bloody minute to listen to the briefing." said Aurora as he desperately rummaged through his packs for a notebook. He also kept muttering something which sounded very much like 'We are all so screwed'.

Hearing Aurora, the rest of the II termers also rushed around feverishly trying to get some writing material to scribble some form of briefing. Aurora looked around and realised the battle was already lost. They had not prepared the range cards[1] or the fixed line[2]. Jay saw what he was looking at and grabbed the range cards. He yelled at Sandy, "Gimme your bloody china graph pencil fast!"

Sandy, who was busy trying to get mud out of his ears with said pencil, realised the gravity of the situation and jumped to it. He threw the pencil at Jay and for once did the sensible thing. He did not crib, cringe or whine. He picked up the LMG and rushed around, waking up the rest of the section. Most repaid this newfound fevour of his with curses and abuses that challenged the legitimacy of his parent's marriage. His response was a sentence compressed into only one word 'Shahcomingbuggerstandto'. Actually five words, but said as one and not misunderstood by anyone.

Aurora and his section managed to get into their trenches, where he loudly and feverishly practised his briefing, till someone told him to shut up so they could think. Actually truth be told, the others could not really figure out what to think. Their minds was blank, first because of the exhaustion, then the rain and cold and finally due to lack of sleep...plus the hearty breakfast. The body can only take so many priorities. But the mind is stronger than the brain especially when it knew the consequences of DS displeasure.

The scene was the same in all the sections with varying degrees of frustration and ambition. The performance of the cadets in the exercises would be crucial to their becoming an appointment in the third term. Jay and Kapoor did not, like many others, realise that their performance and initiative at the camps was vital. These things would be taken into consideration when it came to deciding the cadet appointments. These appointments would help in their position on the merit list and subsequently in getting their choice of arm or service after commissioning. They conferred, came to a quick decision, decided to keep a low profile thinking maybe they wouldn't be noticed. Who cared about appointments on disappointments when your arise was on fire.

It took another two hours before the officers reached Jay's section. Theirs was the second last section. The reason they had to suffer carrying the stone the extra distance the previous day, was now giving them extra time. By this time, except for the vigilant section commander Aurora, everyone had dozed off. He saw the Battalion Commmander's party approaching and ran around trying to rouse his section. He was unsuccessful and unhappy. He would not be able to impress the DSs with his command and control. He would just have to give a damn good briefing.

"GOOD MORNING SIR!" Aurora screamed. The high pitched greeting was a multi-pronged missile. It was aimed at both the officers and the sleeping II termers. It was meant to wake up the sleeping defenders of the section and impress the Bhagat battalion CO. The CO however was in a bad mood after having climbed two hours up the mountain and having listened to ridiculous and at times comical briefings. He was not amused.

The other platoon and company DSs sensed the CO's mood and were ready to inflict punishment. A little bloodletting always lightened any tyrant's or supreme commander's mood. Right now the CO of Bhagat battalion was the king of the mountain and Aurora the cringing serf in front of him.

"All right then start off!" said the CO as the orderly set up the campstools. Capt Shah in the meantime wandered over to the trenches.

Aurora was watching Capt Shah from the corner of his eye and felt a small shiver go down his spine. He started off, "Sir..." and then his mind went blank.

The CO scratched his head and said to Aurora, "Who the hell made you a section commander? Haven't you bloody well prepared?"

Aurora made a massive effort and unfroze his brain. "Sir, if you look to your front you will see a marching line of hackle shaped trees...the centre-most hackle shaped tree...will show the general line of direction..."

"Are you asking me or telling me? Any way which is the centre-most bloody 'hackle-shaped' tree?" asked the CO irritably while the other DSs tittered respectfully in the background.

Aurora managed not to get fazed and continued his briefing, more boldly this time. The others were cheering him on silently without daring to look back. They did not look back because it would have been against good military discipline while listening to a briefing. It was also because nobody wanted to attract any attention.

After two minutes of Aurora waffling, the CO grunted and said, "Ok good... you seem to have made some effort! Now let's go have a look-see at the last section." And with that the CO and his retinue left.

Kapoor was about to heave mighty sigh of relief when he noticed a shadow fall over him. It was Capt. Shah. The sigh turned into a whimper. Capt. Shah asked the dreaded question, "Ok show me the LMG trench and the range cards!"

Jay and Kapoor were totally unprepared and wished they could have used the 'ostrich' defence. They would really have loved to bury their heads inside the trenches. Kapoor decided to man up and jumped into the trench boldly and vaguely pointed out the arc of fire. Capt Shah patiently pointed out the correct way to site an LMG and the arc of fire. The II termers found it all very interesting. Jay and Kapoor found it very soothing. They had been expecting something by way of punitive action.

After ten minutes of Capt. Shah's tuition, the section was feeling sleepy again, and unfortunately Capt. Shah noticed it. "All right the section is now under artillery shelling. Move to an alternate location!" he said and added, "NOW!" when there was no reaction.

The section moved at least three locations with Shah and his imaginary artillery shells following them. The section was reasonably winded if not actually wounded. And it satisfied Shah's standard of tactical training achieved.

'Ok, is there anyone who can actually give a decent briefing?" asked Capt. Shah

To everyone's surprise and Aurora's disgust, Bhasin got up and rattled out a near perfect briefing. Even Capt. Shah was taken aback. "Ok good show!" he said and addressed Aurora, "You can compare notes with this GC next time!"

Bhasin's smirking did not help Aurora's blood pressure any. What came next was worse. Capt. Shah made Bhasin the section commander. Aurora took off the section commander's armband and gave it up like it was the key to his sister's chastity belt. But Shah was not finished, "The rest of you can carry back down that stone you so kindly brought up for me!"

Carrying a load down is more strenuous and nerve-wracking in mountainous terrain. Going downhill in normal circumstances takes a terrible toll of the knees and a heavy burden compounds that. The section discovered this fact the hard way. A smirking Bhasin tried to shirk his part by invoking the perks of section commander. Aurora and the others reminded him that he would also get screwed as section commander if 'his section' did not complete the said task. They managed to reach the base just in time. The other sections had already reported and loaded up.

Capt Shah asked them again, "Hope you guys have learnt an important lesson in the mountains, gentlemen?"

Sandy muttered, "Yeah! Never opt for the bloody infantry!"

Capt. Shah heard it but let it go. But another instructor from the training team did and he was from the infantry. He was hurt by this slur on his parent arm. That hurt feeling resulted in the section having to carry the stone back to the Academy and deliver it to the BCA.

The order was carried out and nobody had the energy to even hate the officer or Sandy. In any case, what Sandy had said was food for thought, deep thought. Joining the infantry meant they would be tramping around mountains most of their career. They had learnt something in this mountain warfare exercise, unfortunately not much of it was tactical. It was more about surviving superiors, but not so much about warfare.

NOTES

1. Range cards-denoted the range to certain landmarks in front of defenses to enable accurate engagement of attackers
2. Fixed line is a frame that has pre aimed slots on which the weapon can be laid to cover certain enemy approaches.

Chapter 29

ACHIEVING GREATER HEIGHTS

Jay ran full speed and vaulted over the wooden horse. The wooden horse was a four-foot high wooden box, about a foot wide and about four feet long. It was a standard piece of equipment in most gymnastic routines. It was not at all what GCs imagined they would be crossing over to become an officer. But the II term physical tests were more about agility and flexibility. The vaulting horse and split horse were designed to test just those attributes.

Jay was among the other II termers who were being given extra coaching by the BUO and a few bored III termers looking for entertainment. The procedure looked simple enough...on the face of it. You took a run up, jumped and landed on the horse with your hands, brought your legs through and landed with your feet together on the other side of the horse.

Sounded really simple and looked that way too when a few of the fitter III termers or PT instructors demonstrated it. It looked real neat on TV too when the gymnasts did it. But most of the cadets there, apart from the military school variety, had never seen or heard of a wooden horse before, leave alone jump over one. The initial practices usually ended badly with one of the two invariable outcomes. The cadet either overshot or chickened out at the last minute. Both outcomes involved crashing, and in spite of the fact there were thick mats there were some minor injuries.

The injuries were not very serious except for one cadet who managed to hurt his neck very badly. Rumour had it he belonged to a rich family and wanted to leave the Academy anyway. The whole horse process was actually quite easy and one just had to overcome one's fear and unleash one's potential. It was however a different thing when one actually tried it and more often than not, it was just fear that was unleashed.

But the PT instructors insisted it was all very simple. But to reiterate, one had to take a jump at least four feet ahead of the horse and literally fly through the air, at least five feet high with legs straight and land with hands exactly on the forward edge. The split horse was equally bad. One had to spread one's legs while jumping and show a correct posture for...well, a split second.

The CSM was watching the proceedings and decided to take a personal interest in the test. The company commander had told him to make sure there was maximum first attempt pass percentage in the PT test. This would count towards the IMA inter-company championship. The CSM then drew a line four feet away from the wooden horse.

"That...you fuckers can consider your '*Lakshman rekha*'[1] or in this case the *Ravan rekha*!" said the CSM who was feeling humorous. Pain, especially others', always made him that way. "The ass who misses this line will feel like Sita." He added unhelpfully.

There were still some cadets who took off after they crossed the line. The CSM and the PT instructor were not happy. They looked at each other and an unspoken understanding immediately flashed between them. Evidently something more drastic had to be done to make the cadets take off from the correct distance.

Unfortunately neither the CSM nor the PT *ustad* were Jedi knights like Yoda. The II termers did not have the 'force', but the CSM was definitely of the 'dark side'. He did what he knew best, use terror, tactically. He decided to use cycles at the jump-off point to force the cadets to take off correctly and jump higher to avoid crashing into the cycles. He was also looking for suitable candidates to make examples of. Unfortunately Jay and Kapoor did not notice the CSM looking at them and were instantly marked as suitable.

Jay remarked to Kapoor, "It's just like the Army to make you jump over a horse!"

"Would you rather a horse jumped over you?" whispered back Jay.

"I didn't know they were going to work me like a dog, that too a bloody circus dog!" replied Kapoor.

"Those two lovers..." said the CSM who had found his targets, pointing at Jay and Kapoor, "stop yapping and go get your bicycles!"

The others relaxed, when someone is told to get their cycle in the Academy, it's not for want of conveyance. It could be considered a garland to felicitate misdeeds and convey a message. The CSM deployed the two bicycles in front of the vaulting horse. Ostensibly that was to add to the challenge, but it only added to the caution.

The PT instructor kept a discreet distance, physically and morally from the proceedings. He did not disapprove of such dangerous coaching techniques, and thus could not be involved officially. For all other purposes, the CSM had the PT *ustad*'s blessings. They too wanted the pass rate to improve, and with that the company ranking for the championship, and ultimately their careers.

The first cadet crashed into the cycles. It took a few minutes to extricate him from the wreckage. The owners of the cycles Jay and Kapoor winced. At this rate their cycles would be trashed and they would have to run to keep up with the others who were mounted for tomorrow's class. The cycles helped improve one's mental and physical agility.

The first victim walked away with some determination and many bruises. The others now faced two choices, jumping as instructed, trusting their inner capabilities and maybe landing badly on the far side. The other option of crashing into the cycles had been demonstrated and proved as painful. Actually there was very little choice. The cycles had sharp ends; spikes, handle bars and other such protrusions, capable of hooking the family jewels. It was better to have leapt and lost than to have landed short and lost a testicle. And this was the general consensus.

Then the CSM said the first useful thing that actually made sense, "Gentlemen it's just this bloody wooden horse between you and third term!"

That one statement galvanised the cadets like nothing else could have. It literally motivated them to greater heights. Even the PT instructor was taken aback by the sudden spurt of '*josh*' displayed by the II termers. But the CSM was not entirely accurate when he had said it was just the horse between them and third term. The third and final term when they would be seniors and lords of all...well, at least the cadets they surveyed.

There was still a mountain between the II termers and the promised land of III term. They still had to cross that. If only they could have leapt over it. They had to complete the final II term camp which even the ex-NDA II termers grudgingly acknowledged may be comparable to 'Camp Rover'. This exercise called 'Silver Chain' entailed crossing the highest peak in that area called Bhadraj. This mountain was not Everest or even close to it, being of a relatively modest height. It was the reputation of the thing and since just as many people had suffered crossing it, it had just as many legends.

The II termers had heard all the legends, and like all II termers before them, had shrugged it off. They could beat it, no problem. The exercise itself was more or less a repeat of their brief mountain warfare sojourn. Jay remembered more about carrying boulders up and down the mountain than anything of tactical value. The experience with the boulder, painful and tendon-wracking as it had been, had nevertheless given Jay and his group a great deal of confidence; in their toughness and ability to take on hardship. The confidence was slightly shaken when they had to listen to the briefing from the BCA.

"Listen up... you lazy bums," he started off mildly, "It's going to very bloody difficult, and some of you will feel the extreme urge to drop out during the final route march! Remember last year at least ten chaps got relegated, because they dropped out of the route march or were medically evacuated. Got it?"

Jay asked Sandy, "Did you get it, asshole? I'm not carrying your sorry ass up the bloody mountain!"

Sandy gave Jay what he thought was a winning smile and simpered. "C'mon yaar, where's your course spirit?"

Jay was about to say 'With your sexy sister' but did not. That might have provoked even a mild mouse with ratty morals like Sandy. The II termers did not really get the gravity of the situation. It's not easy to 'get it' when one is sitting cosy, well fed and full of energy. It is the same as watching a Rambo movie. Easy to watch the hero or heroes fight their way through scores of easily killed villains. It's difficult to do it in real life, as they would find out.

It's difficult to relate to the actual pain and exertion. In the case of the II termers they could relate to the blood sweat and tears, but they felt their previous experiences had made them tougher. They felt confident they could now take on anything. This was what the Army desired, and this sort of feeling would help them carry on throughout their career. Thus, they actually cheered the BCA every time he offered his pearls, nay, bullets of wisdom.

The ACA who had dropped in to meet the BCA had some more wisdom to add. "Gentlemen, at every stage you will feel like giving up...but remember it's just a few hours of pain between you and the best times of your life."

The air was full of cheers and comments like, 'Don't worry sir we can do it' or 'Just make sure you give us a good party when we get back.'

The BCA had the last word, "Guys remember, don't leave anyone behind. Every guy who gets left behind reflects on the integrity of your course, not to mention points for the company championship banner. Also, you may be asked to repeat the runback!" He left them with that sobering thought.

All the II termers turned as one and looked at Sandy and Bhasin who were sleeping beatifically.

"Can we just leave them behind now itself? Please?" appealed Kapoor. The rest of the II termers carried the sleeping pair outside and dumped them in a puddle. Sandy actually took a few seconds to wake up after all that.

"Someone once told me you can't choose your family but you can choose your friends. I guess they forgot to mention bloody *fuddu*[2] type coursemates!" said Jay as they all moved out for dinner. The II termers cheered that as well. A few of them picked up Sandy again and dumped him in the flower bed, amidst further cheers.

The next day, the usual preparations for the exercise started off. The camp stores were collected. Bhasin was promoted to section commander thanks to his performance in the last mountain warfare exercise. The DSs had been impressed with him, unfortunately his coursemates were not. But that's life, the opinion of peers and subordinates are not taken into consideration for promotion to higher rank. The alternative to that practice, unfair as it may sound, would mean anarchy.

Bhasin was finding out just how difficult leadership was when the leader is unpopular. He would have done well to compare notes with Aurora. Actually that would not have been a good idea since Aurora was convinced that Bhasin was trying to compete with him for a III term cadet appointment. Aurora was right. Bhasin had gotten a taste of power and success, just a little taste, but it was enough to get him hooked. That syndrome worked at all levels of power.

Bhasin also realised that in the short term it was important for the leader's bosses to be happy, more than the ones he led. And so Bhasin persevered because even though he did not have most of the qualities of a military, leader he possessed most traits of a political one. Foremost amongst those was a thick skin.

"Hey, help me carry this to my room!" Bhasin said to Kapoor while standing between a pile of camp stores.

The answer was succinct and unprintable involving violent sexual acts that would be physically impossible. It is nonetheless reproduced here succinctly in spirit after some editing, "Fuck off *chutiya*!"

Bhasin took the answer in his stride, he had expected no less. Even Sandy, who was passing by, ignored Bhasin. He had learnt his lesson in the mountain exercise 'Parvat Raksha'. Sandy had personally renamed it to a word in Hindi that roughly translated to sodomy. Bhasin then went looking for some I termers who were not bothered by considerations of good or bad leadership yet, simply because they had no choice.

That evening they had a dinner night. The battalion commander was there. All the officers and cadets were in their mess dress. The mess tables had spotless damask covers and napkins. The crockery and silver was gleaming along with the regimental badges, stars, spurs and medals of the assembly. Jay felt thrilled standing here. His uniform was relatively drab being the winter mess dress or blue patrol, sans any decorations save his term and company badge.

Jay had no embellishments by way of said medals, cross belt or spurs as yet. But just being in this gathering with the officers in their impressive dress uniforms was a kick enough. He almost felt officer-like...well sort of. He envied the officers with the chain mail[3] and spurs, with battle honours on their cross belts. He looked longingly at the missile badges, wings and other such decorations. They made one look so soldierly...and somewhat like a Christmas tree. He felt like a soldier but was not sure he looked like one without the gleaming tokens such as those worn by the instructors.

He felt someone nudging his back. It was Kapoor, "Put your bloody tongue back inside!"

"Come on man! The officers are looking good and I bet if I went back to the cantt. with that kinda dress, the girls will just swoon out of sheer excitement!" said Jay.

"Yeah, the small ones especially...probably out of sheer bloody fright!" retorted Kapoor, "C'mon let's go and talk to Capt Balhara, he must be missing us!"

The GCs had gathered in the anteroom and made polite conversation with the officers. Mostly they gathered around in groups around the officers as they told tall tales. Some of the III termers took the opportunity to suck up to the officers. The merit list was being made and the rating of the third term GCs would make a difference in their ranking in the final merit list. That was all in the hands of the DSs who decided who had the most OLQ (officer-like qualities), hence the importance of impressing the officers.

The merit list would make a difference in their getting arm, corp or regiment of their choice. The sucking up was therefore not officially recommended but considered important. There was no harm in trying and the officers lapped up the adulation. The sucking up was thinly veiled as adulation. As one III termer put it, it was "No skin off his tongue".

Capt.Balhara was holding court with a gaggle of wannabe commandos hanging on to his every word. Balhara, like the average alpha male, loved the hero worship. He was telling his rapt audience some outrageously bloody story of terrorist-cum-insurgent hunting. The story ended with lots of death, bullets and gore which Balhara seemed to find funny, as did some of his audience. There were some pacifists who were a bit scandalised by the scant regard for human suffering or death. Evidently a year at a military academy had still not convinced some of them they were in the wrong profession.

Kapoor asked Jay, "You really think the terrorist blowing himself up by mistake was funny?"

"Not really but I sure as hell don't wanna be Balhara's next target, so please laugh along." replied Jay. He was learning, actually he had learned these tricks watching his father.

The I termers had not yet got their mess dress from the tailor and were wearing their regular working dress. They were a dirty and sad looking lot. This happens with any group at the bottom of the pecking order. Has anyone seen a happy mouse when there are big and then bigger cats around? The officers took pity and tried talking to them and only succeeded in scaring them more. Luckily the BCA called for attention and the battalion commander addressed them.

The battalion commander was positively genial and behaved, for all the world, as if the II termers were his favourite sons, off to the Crusades. He told them pretty much the same things as the BCA and ACA had the previous day. But the important thing was he lectured them in an avuncular tone. Jay realised the closer you got to becoming an officer, the closer you came to being treated like humans and fellow soldiers.

The Battalion Commander did say something significant, "Gentlemen, you have managed to successfully pass your I term and become men. Now it remains to be seen if you can achieve greater heights and become soldiers after climbing the highest peak we have here...Bhadraj !"

There were appreciative oohs and aahs, some manly, some disturbed.

Just then the mess Havildar gave the report[4] to the battalion commander, "*Shriman khana taiyar hai*!" (Gentlemen, food is served)

They then moved to the dining hall and stood behind their chairs while the officers took their place. The I termers had been deliberately placed next to the officers. This was a highly cunning and tactical move to enable the II and third termers to hog without having to bother about etiquette and more importantly, without offending the officers.

The entire dinner night exercise was to teach the cadets how to comport themselves in a formal dinner with many courses and different types of forks, spoons and glasses arranged around the table. Some GCs had come from families that had never eaten at a table; leave alone with a fork and spoon. The Army did not want its officers to make a spectacle of themselves in public, especially in front of the lower ranks. It was thus more of a drill and less about nourishment. But the food would be good, whatever little they got to eat.

Some of the III termers, including appointments had been placed next to the officers. They had no choice, but most of the other III and II termers tried to be out of the line of sight of the officers. That way they could eat in peace and fast. Otherwise they would have to put their forks and knives down every time they took a mouthful, a very small mouthful. They could not keep their elbows on the tables or speak across the table. They had to break only one small piece at a time of the dinner roll on the quarter plate. The list of dos and don'ts were endless. Thus it was nobody wanted to be next to, near, within earshot or in the line of sight of any of the officers. It made the eating experience quite tedious and left one hunger stricken.

The DSs were quite capable of handing out restrictions on the spot for unofficerlike behaviour. The punishment and the procedure were justified but not really fair. After all, these were not officers, just a bunch of well-disciplined and very hungry boys, mostly under the age of twenty-one.

Jay and Kapoor had been seated in front of Capt Balhara. Jay managed to switch seats with a poor I termer just in the nick of time but Kapoor was not so lucky. He had almost managed the switch when Balhara noticed it. He turned to the JUO next to him and said "I don't want to see any piddly bloody I termers near me."

The JUO promptly stood up pointed to some I termers and said, "You, you, you and you, get there, there and there...now bloody move it!"

There was an immediate scramble as the designated 'yous' moved there and here. There was a sense of urgency as nobody wanted the battalion commander to take an interest in the proceedings and escalate matters. The displaced I termers gratefully rushed off to seats far away from Balhara. They stood respectfully and eagerly behind the seats occupied by the soon to be dethroned II termers. One of those seats was held by Aurora. He desperately hissed, "Get the fuck away from me, you idiot!"

But Capt. Balhara noticed this and, said, "What's the matter, Aurora? You don't wanna be near me?"

"Of course not sir...my pleasure sir," mumbled a white-faced Aurora as he realised the evening was screwed.

Jay was also one of those who were caught in the dragnet. The rest of the evening was spent watching and smelling the food. Their mouths watered as they took delicate bites. Balhara probably noticed but he was too busy telling another story about how they had flushed out some terrorists from a house using phosphorus incendiaries. One of the II termers who was a vegetarian, lost his appetite looking at the grilled meat. It reminded him of Balhara's descriptions of the end result of the terrorists.

Balhara took note of this and was mildly concerned about a possible drop in martial spirit. He said, "You seem to be some sissy ordnance or services type optee[5] ?!"

The GCs around Balhara tittered respectfully. It was well known Balhara considered vegetarians sissy and now this. A hungry Aurora in the meantime took his ire out on the I termer sitting across him. He was not supposed to speak across the table but he managed to hiss, "You bloody pig...make a hundred pieces out of your dinner roll and eat it bloody well slowly!"

The poor I termer, from some small town in UP, was overawed by the array of forks and spoons in front of him. He did not really get what Aurora had said but was too terrified to disobey. Sandy, for a change, took pity and helped out the youngster. He told him sotto voce how to use his forks and spoons and in which ordcr. To make sure he did not offend Aurora who was watching like a hungry hawk, Sandy advised the I termer to take very, very miniscule bites. That was what was meant by hundred pieces. It had the same torturous effect as death by a hundred cuts.

The dinner went off fairly peacefully. There was no fall-in afterwards. The GCs went to their rooms quietly to fill their empty bellies with emergency supplies of biscuits. The mess staff settled down to hog on the vast quantity of quality left over's.

But Jay had other plans. He had given a bottle of rum on Diwali to one of the mess waiters. That waiter had come over and taken empty mess tins[6] before the dinner. Hopefully he would bring them back full of food. Jay called over Kapoor and Sandy over to his room. As an afterthought he called Aurora over too, if nothing else but to save some poor I termer from bearing the brunt of Aurora's ravenous wrath.

The waiter delivered the food as hoped. It was all there, the grilled meat, Russian salad, cottage cheese cutlets, scotch eggs; a dozen dinner rolls and last but not the least, tipsy pudding. They switched off the lights to add to the challenge and dug in. They ate everything—pudding, salad, dinner rolls, everything, as one big mess and it was good. They even toasted the Army with some water, same as one did for the President[7] .

Jay had only one condition, "No fucker is going to use or even think about using a bloody spoon or fork! I've had enough of that in the damn dinner night!"

They all belched loudly in accord.

NOTES

1. Refers to the mythical line drawn by Lakshman to protect Sita in the epic Ramayana.
2. Fuddu—slang for 'useless'.
3. Chain mail is worn on the ceremonial or mess dress uniform shoulders by armored corp or cavalry officers. Box Spurs are also traditionally worn by cavalry and sometimes by artillery officers and majors and above of the infantry.
4. In all military messes, the mess-in-charge gives a report when the food is ready or laid out. The format varies slightly depending on regimental traditions.
5. Optee, slang for someone who opts for a particular branch or arm of the Army
6. Mess tins are much like tiffin boxes made of aluminium meant for carrying food based on a design that has seen around a hundred years service.
7. The President of India is always toasted at the end of the dinner night and he/she is always toasted with water, never alcohol of any sort.

Chapter 30

THE END OF THE RAINBOW

A few days later Jay rethought the feeling he had had listening to the battalion commander or batty. He had felt then that he was being treated more humanely as he came closer to becoming a soldier. He was still feeling very soldierly but not feeling particularly well-treated. He supposed the IMA made some distinction between humans and soldiers.

"GCs were not meant to be treated like humans, so that they learn the value of humanity" someone had said. But surely they need not be treated like animals? He was also gradually realising that a soldier in the wild was actually just that. There was no glamour and nobody was feeling particularly officer-like just then. They had been on forced marches the last two days and digging trenches during the night. They hadn't bathed but they had to shave and shaving with little water in cold, dry weather had left them all with peeling jaw lines, apart from their sunburnt noses.

Truly the Academy saying, "You can make a man work like a dog but not the other way around unless it's pussy!" held true. All the II termers of a company were reorganised into platoons and those of the battalion into companies. Thus all II termers of IMA were deployed. The digging and deploying had been hard with the DSs improvising to add to the simulation of 'rigours of battle' hardship. It was a pretty accurate simulation of the hardships of the trench warfare of the Great War aka the World War I, except for the artillery shelling and shrapnel: ...that is if you discounted the rocks occasionally thrown by the DSs to make a point. But the end was near and thus far no one had dropped out. The next day they would have some time off and then the route march back to the Academy over mount Bhadraj would begin.

Jay huddled deeper into the trench and tried to keep himself as warm as he could in his coat parka. He also had his Army issue blanket, but surprisingly even though it was hairy and prickly, it was not warm. It was very much like sleeping with an unfriendly short-haired dog. Most things that were issued in the Army were kind of user-unfriendly anyway. Such were the random thoughts that came to Jay's mind as he drifted off to sleep.

He dreamt he was flying over the big bad mountain on a magic carpet and that too in the dress of an officer. He suddenly heard Sandy's high-pitched voice penetrating the thick fog of exhaustion and sleep. The others grumbled and some yelled at him in various tones and degrees of strong language. The basic theme was the same, "Shut the fuck up and let us sleep or you will get kicked and, or, slapped."

Sandy however persisted, "*Saalon uth jao*...wake up you buggers...the bloody DS is coming!" Sandy meant well. The DS had indeed come. He was Capt. Jeevan and he had been tasked by the battalion commander to go around checking up on the cadets. The cadets were supposed to be alert and manning the defences like there was some real enemy out there. However, they, like generations of cadets before them thought they could outsmart the DSs. Most had gone to sleep, a few after posting lookouts. Also the common thought was 'To hell with the enemy. Let's sleep.'

The lookouts of the neighbouring platoon had also been sleeping and the DS had gone around kicking everyone awake. In one platoon the lookout had desperately tried to buy time for his comrades by challenging the DS and refusing to let him pass without giving the password. It was a gallant if suicidal effort but at least a few of his platoon mates could get back into their trenches, before the DS caught them with their pants down. It was this commotion that had finally woken up Sandy.

Jay, having grown tired of Sandy's wailing, jumped out of the trench to kick him. He saw Capt. Jeevan approaching and jumped back into the trench to retrieve his weapon and hat. He was just in time. Just in time for Capt. Jeevan to see Jay standing resolute in his trench like any dutiful defender was expected to. Standing by his trench would have been better but...Jay was frozen with fear and he was looking at the area in front of his trench and not at Capt. Jeevan because he was too bloody scared.

"Hmmm..." said Jeevan meaningfully. Every meow or purr of the cat is meaningful to the poor mouse. He had had quite a meaningful evening, at least for himself. All the platoons he had checked had either sleeping cadets or sloppy defences. This was the last platoon and there was still another one hour before the officers got together for drinks at the field mess.

Jeevan had uttered his 'hmmm' in reasonably civilised tones. But it had been enough to wake the rest of the platoon up. The II termers had already hatched into soldiers in this term so their instincts were quite well-honed. The combination of Sandy's wailing followed by the sudden silence, woke them up more effectively than any air raid siren. An air raid could be far away but such silence usually meant very near and very clear and present danger.

The 'near, very clear and present danger' in the human form of Capt. Jeevan watched with mild amusement and some irritation as the II termers disgorged themselves from various bushes and holes they had occupied, in a bid to keep warm. He adjusted his fur-lined jacket and smiled. The smile was because he found it funny the way some of the cadets tripped on their blankets as they desperately tried to hunt for their rifles and pack and caps and...well, so on. Jeevan found this sort of thing amusing. He had already been divorced once.

Capt Jeevan also smiled to himself in the anticipation of the fine lesson he was going to teach these miserable excuses for future officers. The lesson being,'the perils of dereliction of duty in the face of the enemy'. The fact that he was the enemy in this case and had also been guilty of a similar crime when he had been a cadet, only made the lesson all the more delicious to dish out.

The II termers shuffled into some semblance of order and stood fidgeting awkwardly, not sure whether to get into the defences or start doing penitent pushups. The smell of guilt and retribution was thick in the air, though Jay and Sandy had a distinct air of righteousness.

This was commended by Capt. Jeevan in his opening statements. "So only two of you fuckers felt the need to man the defences and be on duty!"

The DS's summary of the situation was met with looks that were deeply disapproving and disbelieving. There were indistinguishably mingled with subsidiary emotions like intense dislike and a feeling of betrayal. All these emotions were directed at Jay and Sandy. Some of the worthy defenders fiercely directed all their attention at the area in front of their trench, as if an attack was imminent from that direction. It was a forlorn attempt to escape Jeevan's wrath by showing a belated commitment to duty. Jay and Sandy also looked seriously at the area in front of their trenches. This was kind of ironic since they had plenty of enemies right behind them right then.

Actually, a full-out frontal attack would have been more preferable than the visual daggers their backs could feel from the rest of their coursemates. The coursemates had no cause to feel any resentment. It was nothing personal, just the natural resentment one feels towards someone who has won the lottery.

Jeevan asked cordially, "How is the holiday going so far?"

There was silence. The sarcasm had been duly noted and now that Capt. Jeevan was there everyone knew that the 'honeymoon', 'picnic' and 'holiday' or whatever, was over.

"And where is your worthy platoon commander?" asked Jeevan silkily.

Bhasin who had been appointed 'platoon commander' for the duration of the exercise tried to make himself invisible. To be fair, he had been told someone else would be appointed platoon commander for the route march part of the camp which was the next morning, just few hours away. He pretended he was somewhere else and someone else should, therefore, take the blame. But since that someone had not been named as yet, the crowd parted to reveal Bhasin standing with his head bowed in deep thought.

Capt. Jeevan asked, "So, my dear platoon commander...anything you have to say for yourself... why...why your whole bloody platoon was sleeping...leaving the defences unprotected?"

Bhasin had plenty to say for himself, especially when it came to saving his skin. He jumped to attention and told Jeevan how his platoon refused to listen to him and so on. Bhasin was just beginning to believe his own story when he was rudely interrupted by Jeevan. After all he had a schedule and a colonel to meet with cocktail hour waiting for him at the field mess.

"Alright, the defences were over-run and all of you are dead!" here Capt Jeevan paused for dramatic effect, and continued when only disgusted looks were forthcoming. "In fact, you deserve to be dead and cold...All of you strip into your undies and man the defences till first light...now bloody MOVE!"

The II termers jumped to it. There were no doubts in anyone's mind what would happen if Jolly Jeevan was aggravated. The only ones who were still not very sure were Bhasin, Sandy and Jay. Bhasin felt that as platoon commander he would be excused, and in any case he was convinced he was a victim. Unfortunately for him, Jeevan did not agree with that assessment and ordered him to strip to his undies too and follow him carrying two LMGs. He then turned to the virtuous twosome, Jay and Sandy, still quivering in the trenches and said, "Good show you two... keep it up and don't let up, the duty JCO will be coming to check up!"

Sandy and Jay opened their eyes a full two minutes after Capt Jeevan had left and started breathing a minute after that. They turned around to survey the scene and then rather wished they hadn't. All the trenches had near—naked, shivering defenders with little to defend them against the cold. The looks they gave were pathetic and venomous in turns, but there was nothing to be done for it.

Anyone caught disobeying orders or wearing clothes would be severely punished. It could mean a fate worse than death-relegation. That meant six months more in the Academy as a junior. As one wag put it 'A frozen ass was better than a tattered one!'...that being the loose translation.

Going by the ethos of course spirit, ignoring their coursemates plight would also mean a fate akin to death, ostracism. So Jay and Sandy went around the rest of the night in just their uniform and boots. The rest, cap, gloves, sweater, jacket even handkerchief were distributed and then redistributed to keep some parts of some coursemates warm. The system also ensured the items could be swiftly taken off and thrown back to Jay and Sandy in case of a visit by the duty JCO.

Thus it came to be that Sandy gave Aurora, his nemesis, a pair of gloves, the only other piece of clothing apart from his underwear that he was wearing. The gloves were only symbolic but helped the spirit and the testicles feel warmer as Aurora innovatively used the gloves to cover his family jewels. Sandy did not dare complain as he mentally discarded the gloves or any thoughts of ever using them. Jay and he made some tea furtively, hidden in the bushes, in case the duty JCO spotted the fire. The tea probably helped the defenders survive without serious cold injuries to their extremities.

Luckily the cocktail hour was a reasonable success. Jeevan was feeling mellow after the drinks and reminiscences with the rest of the instructors. His remembering his time in the Academy softened his outlook. He sent back the duty JCO after a couple of hours with the crucial orders restoring clothing. The 'restoring dignity' part would take another ten years.

Jeevan's time in the Academy also told him just how much he could push the cadets before they rebelled or became casualties. By this time everyone was too numb from the cold to kick Bhasin for his incompetence. In any case Bhasin had been sorted out quite thoroughly. He was completely caked with mud and almost staggering with exhaustion. The rest of the II termers quietly crawled into their clothing and then into their defences.

The next day was bright and sunny and apart from the stand to at dawn and rounds by the company commander, there was nothing else to do. They were left alone to thaw for the rest of the day, till lunch time. By 1400 hours, they had filled up their trenches again. There had been a minor hiccup when a local farmer had complained about someone filching fruit and the battalion commander had been displeased with the debris left by the cadets. There had been another round of punishments.

It soon became obvious that the instructors were only doing this to push the II termers to the limits of their endurance. The II termers could smell third term and took all the '*ragda*' without any dropouts. They were too close to give up now. By 1530 hours all blood pressures were back to normal. The II termers were ready to be released for the run back or route march. It was commonly called Bhadraj though the official name was 'Golden Ring'. The unofficial one was 'the burning asshole'. They were ready for both.

Chapter 31

MAN BECOMES SOLDIER

Jay looked at Sandy. Sandy looked back at Jay with what he thought was a sad and pathetic 'look'. The 'look' conveyed sleaze, not pathos but Jay understood the sad cry for help. It wasn't Sandy's fault that he had learnt most of his expressions from his two sisters and one mother.

Actually just about everyone in the company looked at Sandy in turns. Everyone was worried and with a reason. Sandy was one of those sure to drop out. If he dropped out of the route march or runback as it was commonly called, he would probably be relegated and lose a term. That was okay with everyone, except Sandy.

But as someone... probably Shakespeare, or some porn scriptwriter had said, "Therein lies the rub." If Sandy 'dropped out', that is, he refused to run or walk, or fainted or was otherwise indisposed in some remote location, the rest of the company would have to carry him till the nearest checkpoint. They would have to first find the checkpoint through intensive map reading which is never easy in the dark, in a jungle. Then they would be losing time since the runback was timed and they were competing with the other companies, with extra dead weight in the form of a casualty. The instructors would be pissed. It was "a lose, lose situation" as Chib put it.

Aurora hissed, "Ask that bugger to drop out now itself!" He was now the platoon commander and was feeling the weight of responsibility... or was that the tip of the sword of Damocles?

"No way that bugger is going to...he's convinced you are gonna carry him up the bloody mountain!" said Jay.

"Can I wound that bugger slightly with the bayonet or something...?" asked Chib.

"Oh yeah! And what's your excuse gonna be, you dumb shit...accidental discharge of bayonet while weapon cleaning?!" asked Kapoor.

The bugger in question shuffled up to them and said, "Thanks guys...my sisters also said to thank you all for helping me get through the route march...they are waiting in the Academy and have promised to take us out for dinner...tomorrow...if I finish..."

They who were discussing how to take out 'the bugger' now decided... against it.

"No, no, *yaar* what are course mates for huh?" said Chib jovially. He had good reflexes and presence of mind only when women were in play.

The others nodded wearily. Sister or not, he was still their course mate. In fact they were all worried, after all the hype, about crossing the Bhadraj, that too in the middle of winter. They would sweat and when they stopped, they would freeze. Their legs would give out, and last but not least, the ex-NDAs were confident that most of the DEs would drop out. That humiliation would follow them around for the rest of their careers.

The company commander came over to give his combination of encouragement, abuse and other platitudes. Capt. Jeevan stood behind him

and gave a slight nod to cue them whenever the company had to cheer and smiled whenever they were supposed to laugh. Just when they were wondering if the runback wouldn't be less of a pain, the major wound up with, "Gentlemen, leave no one behind and make sure you all get back within the qualifying time or be prepared to repeat it!"

The company commander smiled, but they were not convinced he was joking so they waited for Jeevan's cue. He too hesitated; clearly he was a bit confused. Being a sadist, he found the idea of making people repeat the route march funny but maybe the company commander was being serious. He finally gave a nod and the company cheered. "That's the spirit boys! Don't screw up!" were the major's parting, and not so encouraging shot.

The only one smiling, as they went through the final pack, weapon and control store check...was Sandy. The company commander had added his weight to his sister's charms in the case for dragging Sandy to the finish line. They lined up and waited for the duty officer to say the word. The officer was Balhara, the commando. He considered this kind of route march a routine thing and could not really understand the grim looks.

Capt. Balhara vaguely recalled his time at the Academy and smiled. Still discipline had to be maintained so he straightened the nearest GC's cap and asked them all to warm up with pushups. Some incredulous looks were directed at him, which he ignored. They did seven pushups with packs, one for each loudly shouted letter of the company's name, Singarh. And then they were off on the toughest venture of their lives so far. The first hour went off fine. The company did not lose their way thanks to Chib's initiative in bribing a villager to lead the way. The villager however promptly deserted them the moment it turned dark an hour later. Luckily they could see Bhadraj silhouetted in the sky and headed for it. It was so damned big, you couldn't miss it. But what they did miss was the trail. They tried going through the bushes but after half an hour of trying, they fell back with egos bruised and bodies scratched. Someone suggested a break, a very short one. The time then was 1830 hours. They had to reach the top before 0400 hours.

The short five-minute break extended into ten minutes and nobody moved. Then Sandy got up and said, "C'mon gentlemen let's move it!"

Everybody was shocked into getting up. Sandy showing initiative, especially for physical exertion, was unheard of. As they moved around adjusting their packs and checking their stuff, Chib asked, "What the fuck is wrong with

Sandy? I have a feeling his sisters are not gonna be waiting for us at the Academy!"

"Relax dickhead, he wants all of us to be fresh and able to carry him...also if we all fail he fails too...he's just bloody concerned about us!" said Aurora who was feeling a bit miffed for not having cut short the break himself.

They scouted around some more and then they heard the breaking of bushes, people falling and a whole lot of cursing. It was another battalion's II termers. They all went to the ground as one and waited for the other company to pass by. They did and they were seemingly on the right track. They waited for a few minutes and then followed the other company. The other company seemed to be on steroids or maybe they had a good guide because they soon vanished. But that did not really matter as Singarh Company had found itself back on track, literally.

Another half an hour and they were at the base of Bhadraj. Everyone looked at the massive mountain that seemed to block out the moonlight, the stars and the horizon, and all chances of getting back to their now beloved Academy. Sandy and a few others at this stage started showing their true colours. Someone said, "Let's have food." They were referring to the packed food they had been issued. This 'packed dinner' was probably a mashed homogenous mass that had reverted to its precooked self.

Aurora sensibly advised against it. His reasons were sound; it was difficult to walk on a full stomach. Time would be wasted and then more time would be wasted when the exertions made you vomit. Someone famously said "The masses are asses"...he should have added, tired asses are hard to reason with. Another ten minutes were wasted while a leisurely meal was had. Jay and the others just nibbled at a *puri* and drank sparingly. The next water point and medical aid post was six hours up the mountain. They had already wasted about two-and-a-half hours going around in circles.

They started moving up the mountain. The track was quite good. That is if you were a mountain goat. At that point most of the humans on that mountain track were finding it difficult to understand the goat's point of view. After an hour of climbing the first II termer sat down. Surprisingly, it wasn't Sandy but some ex-NDA. "I can't go on, man. I'm exhausted...help me!" said the cadet named Prasad.

Kapoor laughed. It was a meaningful laugh, however a bit out of context. You see he had copied it from some bitchy soap opera mother-in-law. It was supposed to convey scorn, probably like the mother in law felt. But he was exhausted and feeling more like the daughter-in-law so ended up sounding like a screwed up goat... no, not the mountain one.

Aurora however got the meaning of that laugh and was outraged that his beloved NDA had been let down "that too in front of these bloody DEs". "C'mon guys let's help him...I got his pack. Someone get his bloody rifle!" he said to the other II termers, preferably ex-NDAs.

No one got anything; they just got up and started moving. They had reached that state of mental and physical exhaustion where it was all about survival. Jay walked up and picked up the rifle and someone else dragged Prasad up. Aurora was mortified that none of the ex-NDAs had come to help. Seeing Aurora's disappointed expression, Kapoor summed it up, "Fuck it man, we're all in the same bloody boat, let's just keep moving!"

Another two hours and they took a break. They just had to. The break was kind of involuntary. The dinner mongers had started vomiting and many others were getting cramps in their thighs. Aurora was struggling on gamely but the double pack was straining even his prodigious stamina. Then someone offered to take Prasad's pack. Aurora kicked Prasad and told him, "At least carry your own damn rifle you asshole...you're a soldier now and soldiers don't make others carry their weapon!"

That had the effect, and Prasad grudgingly retrieved his rifle from Jay. Jay looked around and in the moonlight he could make out a flurry of activity as some went around trying to get back their rifles. They had heard Aurora and had been shamed into making an effort.

The group was divided into the 'haves and the have-nots' and it was not about stamina. This forced march was not about stamina but about mental toughness and endurance. Some just did not have the will to carry on. So others carried their packs in spite of being strained to the limit of their endurance.

They started off again, with the pattern of walking for a while till they had to stop because someone had cramps or refused to go on. The malingerers would be kicked and cajoled till they got up and the cycle would start again. Some of them had genuine problems; some just didn't care or had reached the end of their endurance. It was a very steep climb in the middle of winter. They were just not used to it. A trained soldier would have found it just as tough but would have known that he could do it as he had done worse. That was what this route march hoped to achieve. But not everyone had the heart of a soldier...as yet.

Jay did not feel much like a soldier but just kept his head down, put one step in front of the other and carried on. His legs felt like they were on fire. He was already carrying two rifles. One of which belonged to Sandy. Kapoor was carrying his packs. Jay was wondering how long they could carry on when Aurora shouted out, "Gentlemen, there's the top!"

Everyone looked and in the distance they could see a light flickering. That had to be the summit where the next checkpoint and medical aid post would be. The entire group was energised. They croaked out a cheer and picked up their asses literally and figuratively, and headed for the light...of their life...literally and figuratively.

After 40 minutes of spirited hiking they reached the light. It was a shepherd's cottage, more of a shelter for sheep and goats than for humans. Indeed the shepherd was sleeping by an open fire. He was now watching the rag tag group as they gathered around him. They looked kind of disappointed.

"Checkpoint *kahan hai*?" asked Sandy. Aurora gave him a look reserved for dirty vermin; Sandy had beaten him to the punch again.

"Checkpoint *kahan hai*?" echoed the shepherd. He saw this kind of creature every six months, but no one had asked him where the checkpoint was...whatever that was.

Bhasin asked eagerly, "Let me interrogate him! I'm sure he knows!"

"You think he has any sisters...?" asked Chib of Jay, in sotto voce. Exhaustion had caused Chib's brain to reboot itself to its primeval settings. Jay gave him Aurora's 'vermin' look and before he could, Aurora said it for him.

"Shut the fuck up, will you?! We are not in bloody counter insurgency, Bhasin you idiot!" said Aurora and then turned to the shepherd and asked if he knew of any army presence.

The shepherd could not understand what was being discussed but survival instincts told him the sooner he dispatched this band of dirty and desperate looking brothers, the better. He had earlier seen some uniformed men going up with mules laden with stores so he simply pointed up the track and said "*bas pass mein.*"

The group was not encouraged by the shepherds answer. It was a well know fact that any highlander if asked for directions would give an encouraging answer like 'close by' or 'just one hour'. It did not matter how far or bad the route to the objective actually was. The majority sat down and said as one 'BREAK'. Aurora had no choice but to concede.

They were an hour behind their ideal time and he was worried their timing or the lack of it thereof would affect his chances of getting an appointment in the final term. After a few minutes, he got up and found the majority had curled up around the fire and gone to sleep. Some had, depending on preferences curled up around the sheep. Aurora sighed...he preferred goats himself, but time was getting on.

"Get up you buggers!" Aurora said mildly...actually it was tiredly.

Gradually the II termers got up and checked their weapons and control stores again automatically. Aurora said "Gentlemen just one more fucking hour, so let's get a move on and bloody finish it!"

The others, groggy from sleep and exhaustion echoed weakly 'Yeah...yeah! Let's bloody finish this damned thing'. With this they moved onto the track and started off again still chanting, 'One more bloody hour...'

It took another hour as the group of exhausted II termers from Singarh company stumbled and clawed their way to the top. They saw a fire, with some officers dressed warmly, being served some hot steaming drink. The II termers were mesmerised by the sight and thought as one that all this would be theirs...someday...if they made it down the damned mountain. They headed straight for the hot cocoa in the flasks. Another company had just finished and were getting ready to give their report and move down the mountain.

The first few Singarh II termers had just started to pour themselves cocoa when they heard Capt. Jeevan's distinctive snarl, "Gentlemen, don't behave like a fucking civilian mob!"

That was enough to bring them back to the military thought process. They automatically fell into three lines and started checking their weapons and equipment. Aurora took the headcount and then made the dreaded statement, "There's one short!" This was followed by a hushed silence as various possibilities and their ramifications were digested.

Jeevan, who had overheard Aurora, walked up and hissed into his ear, "Find out who the fuck is missing and find him...nobody is bloody moving from here till you give me the all ok report!"

Aurora looked around helplessly. Everyone averted their gaze and some started moving towards the cocoa. They obviously felt it wasn't their problem. A situation like this was actually everyone's problem, but it was up to the leader to provide a solution. The leader Aurora, was looking lost. He was, after all, human, and exhaustion does slow down the senses. Jay and Kapoor in the meantime realised the missing person was Sandy. They would not have classified him as a person at that stage. They remembered the last time they had seen him was by the shepherd's fire.

Jay, Aurora and Kapoor threw down their packs and rifles and told Chib to keep an eye on them. They then rushed back down to the shepherd shelter. They were sure they would find Sandy there. They were exhausted but running on adrenaline and reached the shepherd's place in half an hour.

They found Sandy there curled up by the fire, inside the shepherd's filthy, smelly quilt with a beatific smile on his face. Aurora picked up a stone and had to be restrained by Kapoor. Jay picked up a stick and whipped Sandy. But Sandy was too deep in slumber, dreaming of his mother's lap and protected by the thick quilt, to be woken.

Then Kapoor and Aurora took off their web belts, pulled off the quilt and laid into Sandy. It still took Sandy one full minute to come to. He opened his eyes and saw the heavily breathing trio and yelped. This time Aurora caught Kapoor who was drawing back his heavy boot for a kick that would have taken care of Sandy's family planning for some time to come.

"No! We have to get this bastard back and give the report!" said Aurora who was beginning to regain his leadership.

"Guys I'm really sorry...it was just for a few minutes..." whined Sandy.

He got a kick and a shove in reply. No one had the breath to tell him two hours would be wasted, thanks to his nap. They had moved a few yards when Sandy ran back to his previous resting place. He had forgotten his rifle. The others rolled their eyes and prayed silently for rescue from future officers such as Sandy and hoped he hadn't lost anything like a bayonet or ammo. "This fucker will someday make history by being the first to get fragged[1] in the Academy itself!" said Kapoor.

They reached after another forty-five lung searing minutes. They were greeted by sight of well rested and cosy Bhasin who asked petulantly, "What took you guys so long?!" He was slapped across the face by Kapoor and skulked back into the dark. Aurora got the rest of the company fallen in and gave Jeevan the report. Jeevan was suitably impressed and just grunted, "Good! Now push off...you guys are an hour behind!"

Sandy asked, "Guys, can we take a break and have some cocoa?!"

Nobody had the strength to say anything. Jay's shirt was soaked with sweat and he was freezing, so he took out his IMA T-shirt and wore it under his jersey. He then grabbed Sandy by the collar and they moved off to join the rest of the company. The move down the hill was just as exhausting and it took a terrible toll of the knees. There were a number of falls and tumbles which necessitated frequent checks of the weapons and stores.

They reached the bottom just as the first rays of light appeared over the mountain. Aurora called for a break. "Gentlemen kindly check your bloody stuff...I will fuck any bloody asshole who later on tells me he's lost some control store!"

There were a few smiles as they could now see light at the end of the tunnel. Aurora continued, "Guys we are just about twenty kilometres from the Academy and just two more hour to make it or be disqualified...so please let's bloody do it! And...try and help each other!"

Sandy nodded eagerly when he heard the last line. They finished whatever water they had and shared what chocolate or food they had and moved on after a raggedy cheer. They moved at a slow shuffling jog which degenerated into a fast stagger every five hundred yards or so. After covering half the distance they took a break. One of the II termers fainted and was revived. He gamely stood up, fell down again and insisted on carrying his own pack and rifle. He fell down again and got up on his own and carried on, supported on both sides. Sandy started crying and said, "I can't go on," or words to that effect. No one had any energy or sympathy to waste and they just walked by him. This close to the Academy, even the instructors would not blame the company for leaving behind a shago like Sandy.

Sandy wept bitterly. Jay asked Chib, "Aren't you going to help him?! At least...for his sisters' sake?"

"Fuck them!" was the tired answer.

Still Jay gave Chib, Sandy's pack, picked up his rifle and kicked him to his feet. They moved slowly because of the need to keep kicking a sniffling Sandy forward.

They finally reached a bridge about a kilometre from the Academy and the finish point. Sandy stopped his sniffles and asked, "How far is the finish line from here?"

"Less than a kilometre." answered Chib.

Sandy was suddenly galvanised, "Gimme my rifle...a soldier should always carry his rifle..." With that Sandy grabbed his rifle from Jay and moved off at a brisk pace towards the Academy.

Kapoor yelled after him, "Take your pack also motherfucker!" But Sandy had already moved out of earshot.

Jay and Chib followed, carrying Sandy's pack and other equipment. The reached just a few steps behind Sandy and were met by the battalion commander who beamed at Sandy then turned to Jay and Chib and said, "Hurry up gentlemen, don't be laggards..."

Chib looked at Jay and said, "For that Sandy owes me his sisters and his mother..."

Jay looked at the rest of his coursemates spread out around the ground. They were hollow cheeked with sunken eyes, some still bleeding from scratches and bruises, but they had recovered their spirits. Some were even laughing and joking around. There were no traces of the desperation and exhaustion of the previous night. Jay picked up his pack and went back to his cabin. The I and III termers had made collages from old magazines and stuck it up on their doors. They had also stacked sumptuous snacks and drinks in the middle of the quadrangle. The big banner said 'Welcome back soldiers'.

Some of the I termers asked Jay, "Sir, how was it?"

"That's not the question you ask..."murmured Jay.

"It was tough!" said Sandy who seemed to have recovered fully "Only real soldiers could have gotten across that damned mountain!"

'Yes', thought Jay...they had finally become soldiers. Someday he would become an officer. Someday...someday soon...But now they were going to III term, the final term...he would be junior no more.

NOTE

1. Fragged, a term which came into popular usage during the US Vietnam conflict. It refers to the practice of soldiers killing unpopular officers.

HINDUSTANI WORDS AND TERMS

Aaja	Come
Accha	Good
Ajgar	Python
Baap	Father
Baki	The rest
Beta	Son
Bhosadike	Slang, form of abuse refers to testicles
Chapattis	Unleavened bread, widely consumed in South Asia
Chot	Hurt
Chutiya	Slang form of mild abuse
Dadaji	Paternal grandfather
Dandaman	Literally stick man, used for banging on the vehicle tail board to make it stop
Dekhna	To see, view
Denge	Give
Dhakka	Push, shove, also a nickname for DEs
Dil ka bhoj	Burden of the heart
Fauji	Military person
Fuddu	Slang abuse equivalent of 'pussy'
Gaya	Gone
Ghoda	Horse
Guru Shishya	Teacher Student

Hai	Is, have
Halka	Lighter
Hammam	Hot water boiler
Havildar	Sergeant
Hoga,	Happen
Homopanthi	Slang for homosexual activity
Hota hai	It happens
hota hain	It happen
Hum	Us
humne	We
Jaldi,	Faster
Jemadar	Above NCO equivalent to Warrant Officer
Jhatkas	Jerk, shake, also a nickname for OTA cadets as they complete their training in about nine months
Junglee	Forest dweller or uncivilised
Kaam	Work, task
Kaatega	Bite, cut
Kadak	Strict, still
Kala	Black
Ke liye	For
Khadus	Cranky
Kharab	Bad, Spoil
Kiya	Did
Ko	For
Langar	Communal kitchen
Mar jaega	Will die
Maroed'	Slang for stolen or pilfered
Masti	Excitement
Mein	Myself, I
Nahi	No
Nahin	No

Nanga	Rude
Padhne	Study
Pasina	Sweat
Pracharak	Worker, volunteer
Ragda	Scrubbing, also slang for harsh punishment or hazing
Rakab	Stirrups
Rehna	Stay, keep on
Sardar	Leader, also slang for Sikhs
Savdhan	Attention
Shers	Lions / Tiger also used to describe Urdu couplets
Siraf	Thats all
suna	Heard
Use'	Him /Her
Ustads.	Masters, Teachers, NCOs are usually addressed as such
Uth jao	Getup
Yaar	Friend
Yaar	Friend
Ziyada	More

Common Military Terminology and Slang

ACA (Academy Cadet Adjutant)—One of the few appointments to have a phone in his cabin. Took orders directly from the academy adjutant and passed them on to the BCAs.

ACC (Army Cadet College)—Once only co-located, but now an integral part of the IMA where cadets who have been selected from the ranks are sent for pre IMA training.

ACR (Annual Confidential Report)—A performance appraisal written for all officers annually and taken into consideration for promotion.

ADC (Aide de Camp)—Authorized to all two star general officers and above while in command.

Adjutant—The key staff officer in charge of coordination and discipline in any military unit especially the Academy. Usually a highly decorated or upcoming officer.

Adm—Short for administration and pronounced as 'adam'.

AEC—Army Education Corps.

AMC—Army Medical Corps.

AOC—Army Ordinance Corps.

Appointments—Refers to the cadet appointments or office holders.

Arc of fire—The area that one can cover with a weapon is called arc of fire.

ASC—Army Supply Corps.

Bajri order—Packs loaded with gravel was called that.

Battalion Commander—Usually a Colonel.

Battalion or paltan—As it is commonly referred to in South Asia has four to five companies depending on its role. Artillery and armoured units are referred to as regiments.

Battle Honours—Units that have performed well in any battle are awarded the name of that battle as an honour. Example 'Shamsher Nagar' after a locality in Bangladesh during 1971 war, 'Cassino' after Monte Cassino in Italy which was a particularly hard fought battle during the World War II.

BCA (Battalion Cadet Adjutant)—At battalion level, took/ relayed orders from the ACA to the CSMs.

Bhagat Battalion—A training battalion in IMA, these were named after famous generals.

BRO (Battalion Routine Orders)—Give out among other things the detailment of duty officers and JCOs/NCOs, etc. The Academy BRO also gave out the various types of punishments awarded to the cadets with the reason and quantity.

Buddy Pair—Men are detailed in pairs, so at any time everyone is accounted for.

BUO (Battalion Under Officer)—The cadet appointment in charge of a battalion. The person would be formally addressed as Gentleman Cadet Battalion Under Officer so and so of Bhagat Battalion but officers and juniors in the academy would call them just BUO.

Cabin cupboards, line inspections—The equipment, bed, furniture and other belongings are kept neatly in a prescribed manner ready for inspection.

Cantt. (or short for cantonments)—The common term for military bases.

Cavalry—The Armoured Corps has tanks now but is sometimes referred to by their old name 'the cavalry'. There are only two horsed cavalry units in India 61 Cavalry and the PBG (President's Body Guard).

CDS (Combined Defence Services) exam, for joining IMA or NDA, but held on different dates with different papers/subjects.

CO (Commanding Officer).

Col—The official abbreviation for colonel. Lieutenant Colonels (Lt. Col.) are also generally addressed as colonels.

Command Mess—The Command HQ officers mess, all units and HQs with a certain strength of officers are authorized a mess.

Commission—Graduating from the academy and becoming an officer was known as earning your commission.

Control Store—Almost all military hardware comes under control store, but in this case refers to items that are hard to come by like bullets, compasses, maps, etc.

Corp Commander—Commands a corps which usually comprises three divisions or about 40-50000 troops.

Course Spirit—Cadets belonging to a particular batch or course were expected to be loyal and display solidarity even in the face of extreme adversity to members of the same course. The spirit and bond carries on well beyond even retirement.

Cpl—Corporal being the full form—a two striper called nayak in South Asia.

Crawling—Another method for moving with a weapon. Could be painful if made to do it on a road especially if only elbows were to be used.

Cross Belt—A belt worn across the shoulder with decorations, worn on ceremonial occasions by the armoured corps and light infantry.

CSM (Cadet Sergeant Major) in charge of discipline and passing orders, a major lynch pin in all armies.

CUO—Company Under Officer.

Cut Surd/mechanized surd—The common slang for Sikhs who have cut their hair and trimmed their beards, could also refer to those who had trimmed their hair and shaved their beards.

DE (Direct Entry)—This being the type of entry into the IMA after graduation from college with a degree and qualifying in the CDS plus SSB. The other cadets come through NDA and ACC.

DMO/RMO (Duty Medical Officer) found in hospitals and Regimental Medical Officer is the doctor posted to the unit.

Drill Instructors—Their main focus was coaching (putting it mildly) in drill, but they were also used extensively to maintain discipline and decorum in the Academy.

DS (Directing Staff)—The common term to indicate officer instructors.

ED (Extra Drill)—Another official punishment, consisted of a hectic session of drill, after lunch that left on quite...moved.

Ex PPG—Excused PT, Parade, Games.

Exercise—Generally indicates tactical exercises or simulated battle action.

Fall in/out—Getting into a parade of assembling was falling in and breaking off or dispersing (always in an orderly fashion unless told to get lost and fast) was falling out.

Field Mess—Officers mess assembled in the field, usually a modified tent, can be quite fancy.

Fixed line—These are pre aimed lines fixed for laying the weapon and firing on likely enemy approaches even without adequate lighting.

Flat Foot—One of the warming up techniques for PT and drill, it involved jumping till your knees.

GC (Gentleman Cadet)—As the officer trainees of IMA and OTA are referred to. The ones in NDA and ACC are just plain cadets.

Grenadiers—An infantry regiment, historically had troops with a higher height average and athletic build to help them throw grenades further. They have a large proportion of troops from Rajasthan as well as Ahirs and Jats.

GTO (Group Testing Officer) in the SSB, usually of the rank of Maj./ Lt.Col.

Guards—A famous regiment that has some of the oldest battalions in the infantry, formed after independence, now mechanized infantry.

Gurkha Regiment—One of the many infantry regiments. There are many Gurkha regiments, each with their own unique number and class composition.

Hands Down—Getting into push up position.

Havildar—Sergeant.

HOTT—Head of Training Team.

IMA (Indian Military Academy) Dehradun—The oldest of the officer training academies. The graduates get permanent commission with a minimum service bond of 20 years.

Infantry Brigade—A brigade the lowest fighting formation with usually three units with other support corps sub units like Signals, ASC etc. An infantry brigade would have infantry units and artillery brigade artillery units and so on.

JCO (Junior Commissioned Officer)—The successor of VCOs, They wear a regimental coloured ribbon and one star along with regimental flashes (akin to a sub-inspectors rank badge and often mistaken for one).

Jemadar—A JCO equivalent to Warrant officer above NCO's and below Officers, this particular term for JCOs is found these days in the armoured Corps and Cavalry.

JUO (Junior Under Officer) in charge of a platoon.

Lancee—Slang for Lance Corporal wears a single stripe.

Landmark—While indicating targets or briefing on takes the help of landmarks and often give exotic names like 'hackle tree' or butterfly bush for instance.

LecDem /LDM—Short for lecture demonstration.

LMG—Light Machine Gun.

Lt. (Lieutenant), wears two stars.

MCO (Movement Control Office)—Commonly found at almost all railway stations. Assist military personnel in transit and coordinate movement of units.

Mess Dress—Differed based on regimental or corps pattern and time of the year. Winters blue tunic and trousers and in summers white shirt and cummerbund with black pant. Both had regimental flashes, badges of rank and shoes or Wellington boots.

Mess Hav (Mess Havildar)—The NCO in charge of the mess.

MH—Military Hospital.

Missile badge—One is eligible to wear a missile badge on successful firing of an anti tank missile, worn above the right breast pocket, the shape would depend on the type of missile fired.

Musketry—PT with weapons, meant to strengthen the muscles used for firing.

Mutual Support—Trenches and defences are sited in a manner so as to provide mutual support.

NA (Nursing Assistant)—Found in the AMC, quite a cheeky lot.

NCO (Non Commissioned Officer) traditionally wear stripes on their uniform sleeves in most armies the world over.

NDA (National Defence Academy) in Khadavasla near the city of Pune (Maharashtra) one can join after graduating high school.

NDA Blazer—This blazer was quite distinctive with the tri-service NDA badge in gold embroidered on the breast pocket. Many from NDA continued to wear it with pride even though they were issued iron grey woolen suits in IMA.

Nk—Abbreviation for 'naik' or corporal, wears two stripes.

OLQ (Officer Like Qualities)—A set of laid down characteristics expected in an officer. The more such characteristics shown the better the officer material. Used to assess candidates and cadets.

Orderlies—The batmen or soldier servants were sometimes referred to as orderlies.

OTA (Officers Training Academy) in Chennai (Tamilnadu) with another being opened recently in Gaya (Bihar).

Outpass—In the NDA its referred to as 'liberty' and in the IMA outpass it's basically permission to go into town with restrictions on which parts that could be visited. The dress code was strictly enforced and those on punishments usually could not go.

Pack 08—A backpack introduced in 1908 and removed from service only in the mid nineties.

Parade State—This was the official details of how many personnel were on parade or present, along with the details of those present and or on duty and why. A major part of any CSM or adjutant's time went in either making or verifying the parade state.

Paras—The common slang while referring to the Parachute Regiment.

PBG—Presidents Body Guard.

Platoon—This is part of a company and has three sections.

PT—Physical training.

Quadrangle—The cabins of GCs in IMA were arranged in a square with a small lawn and covered parking spot for cycles.

Quick march—A quicker speed of marching used for warming up.

Range Cards—The card made with landmarks showing ranges as well as arc of fire and fixed lines.

Restriction—Term for the official punishment, consisted of a five km run, a report in the afternoon and then the night at the drill square where the drill instructors 'warmed' you up before taking the report.

Risaldar—A JCO rank, equivalent to Subedar, still used in the armoured corps and cavalry unit.

RMO—Regimental Medical Officer.

Rolling—Correct term being front rolling wherein one curled up and rolled like a ball mostly on your back. Supposed to help you move low and fast while under fire but used as a punishment. Had the double effect of bruising the back and causing giddiness for beginners.

Route march—Long forced marches with packs and sometimes weapons, gets easier with practice.

RVC—Rashtriya Veterinary Corps.

Sahib—Officers and JCOs are addressed as sahib by NCOs and sepoys.

Sainik Schools—One per state, with five military schools earlier known as King George Military Schools and one RIMC (Rashtriya Indian Military College) all with the prime aim of grooming their students for the military and succeeding spectacularly.

Sawdhan—Attention as in getting into attention.

Section post—The trenches or defences of a section part of a platoon.

Shago—Weakling-one who is shagged out.

Shaktiman—A truck with a three ton payload manufactured by the Ordinance factories, currently obsolete and phased out.

Shoulder board—The cadet appointments wore shoulder boards with red colour for company/battalion and black for academy appointments.

Sick Report—Those reporting sick are assembled and checked by the duty officer or JCO and then marched to the hospital by the duty NCO. Vehicle is provided only if too sick to walk. One had to be very sick to get that.

Singarh Company—One of the GC companies of IMA named after the battle of Singarh. These companies were named after famous battles and victories by the Indian army or historically significant.

SIQ (Sick In Quarters)—Given when the person is too sick to attend parade but not sick enough to be admitted.

SM (Subedar Major)—The senior most and highest ranking JCO, considered a lynch pin of any unit.

Squadron—A subunit like a company or battery, found in the air force. Also training companies in NDA are called squadrons.

SSB (Services Selection Board) for all officer candidates.

Subedar—A JCO wears two stars along with regimental coloured ribbon on his shoulder. Senior to Naib Subedar with one star.

Understudy—All cadet appointments were allowed to have an understudy to help prepare the parade states and generally be an unofficial flunkey.

Ustaad—NCOs are referred to as ustaads.

VCO (Viceroy Commissioned Officers), created during the British raj and received the Viceroys's commission as opposed to King's commission for officers. Supposed to be a link between officers and the men.

Web Equipment, The packs, ammunition pouches, anklets (worn over boot tops to secure the trouser legs) and belt were made of a tough canvas weave known as webbing, also because it crisscrossed the body like a web.